The Other Side of Safety

The problem with the way the safety industry functions is three-fold: (1) the dysfunctional relationship between business and safety leaders, (2) the practice of Results-Based Safety, and (3) the creation of a false reality. This book presents an insightful and practical approach to how you can move your safety program from Results-Based to Behavior-Based Safety.

The move involves understanding what motivates behavior, utilization of consequences, practicing the seven steps of performance coaching, creating accurate safety campaigns, and defining evidence of a healthy Behavior-Based Safety program—this is the other side of safety.

The text:

- Defines the four major motivations and explains how they work and how safety leaders can use the right motivation for the right person to help them practice safe behavior.
- Explains how to maximize the impact of reinforcement consequences and minimize punitive consequences in a way that is aligned with an individual's motivation
- Implements the seven steps of performance coaching conversations and how safety and business leaders can model fluency and frequency to shape behavior to habit strength
- Provides clearly defined evidence of a healthy Behavior-Based Safety program by measuring outcomes like locus of control, self-esteem, self-efficacy, and self-actualization
- Highlights the distinction between Results-Based Safety (RBS) anecdotal practices and the science of Behavior-Based Safety (BBS) methodology
- Showcases how the distinct difference between a mechanistic and organic culture and how the four phenomena can be utilized to drive a safety culture on purpose
- Discusses the importance of expanding from lagging indicators to leading indicators for robust metrics and predictability
- Addresses the significant negative impact of "telling people what to do" and re-focuses on coaching people on "what to think"

The book provides definitions, examples, and applications that focus on how safety and business leaders can influence the behavior of people, impact their culture, and support healthy relationships. It will serve as an ideal text for students, professionals, and researchers in the fields of ergonomics, human factors, human-computer interaction, industrial-organizational psychology, and computer engineering.

The Other Side of Safety

Moving from Results-Based to Behavior-Based Safety

Robert Palmer, PhD

CRC Press is an imprint of the
Taylor & Francis Group, an **informa** business

First edition published 2023
by CRC Press
6000 Broken Sound Parkway NW, Suite 300, Boca Raton, FL 33487-2742

and by CRC Press
4 Park Square, Milton Park, Abingdon, Oxon, OX14 4RN

CRC Press is an imprint of Taylor & Francis Group, LLC

ISBN: 9781032365565 (hbk)
ISBN: 9781032375601 (pbk)
ISBN: 9781003340799 (ebk)

DOI: 10.1201/9781003340799

Typeset in Sabon
by Deanta Global Publishing Services, Chennai, India

Contents

PART 2
Applying the science of behavior

PART 3
Structuring the culture for functional safety

Preface

This book was written as a result of research. My focus is to help organizations and industries improve safety operations through the performance of people. The intended audience is business leaders and safety leaders who, through their position, and ability to influence/affect decision-making and safety practices, create safety culture and have an impact on how the organization influences their people to fully function to be their best.

The focus of this book is far reaching and its audience extends from frontline safety people, safety managers, senior safety leaders, to business leaders who influence the conditions of the safety culture, the style and tone of communication, demonstrate attitude, determine how behavior will be responded to with reinforcement or treated with punitive reactions, how decisions get made, and with whom the organization will operate via financials, procurements, contracting, and selection of personnel.

Traditional safety is really Results-Based Safety. Led by safety organizations like OSHA, IOSH, NEBOSH, Safe work Australia, British Safety Council, Canadian Center for Occupational Health and Safety, Health and Safety Authority, and Korea Occupational Safety and Health Agency, business and safety leaders have become fully focused on safety results, and they have confused Results-Based Safety (RBS) with Behavior-Based Safety (BBS). Results are an indicator of whether you are effectively collaborating, training, educating, and conversing with your workers (i.e., leading, coaching, and regulating effectively). Generations of safety leaders have practiced Results-Based Safety viciously. Business and safety leaders can be mean spirited, arrogant, condescending, express visions of grandeur, create safety cultures of fear, and the worst thing ever is that they never really care for their people properly, just the results. The world of safety can be very abusive, like bad parents who are dysfunctional, but the business and safety leaders practice denial because deep down they don't believe in people. Safety managers are forced to create an environment of chronic dis-ease, keeping people in fear, anxious, stressed out, and worn out by constantly focusing on how they can ensure that their company operates safely. This is not a functional, healthy, or sustainable approach to managing the operational risks in business.

Business and safety leaders face ever-increasing expectations and obligations placed on them. The world is infatuated with results, and the world of safety is the same, and zero accidents is the ultimate achievement. However, how zero is achieved has become so dysfunctional that serious safety leaders argue with safety experts and with their people on which dysfunctional practice is better than the other. It's not healthy when safety organizations like OSHA or business leaders and safety leaders use the threat of punitives (i.e., punishment, penalty, and extinction) to get results. The legal community also makes it tough with laws that are solely punitive and drive fear into the hearts of business and safety leaders when mistakes are made. Don't get me wrong, we need safety law and order. The core problem is that you can't lead people to fully function and perform at their best based on fear. Fear causes people to react rather than respond, and that's a bad thing. Professional people are trained to respond. In psychology, we call this the fight or flight response. The safety community, unfortunately, is driven by fear and continues to function in dysfunction, driving flight (hide injuries) and fight (passive-aggressive attitudes) in people. The safety community needs to change from fear to passion and recognize how to work with each other in a way that motivates people to fight for safe behavior every day. Unfortunately, the tradition in safety has been flawed from the beginning and will never create sustainable results because results are the wrong focus.

Results-Based Safety is rooted in bad leadership. Business and safety leaders have this misguided view that they make the important decisions, they *tell* people what to do, and the workers need to follow the great leader, and if they do, results will be achieved. If they don't do as they are told, then there are problems, failures, pressure, and stress at extreme levels—and leaders see their role as battling these people and getting things done. In the life of a leader, they think to themselves, no one really understands what it's like at the top, "doing what I have to do, and what I put up with." This is all wrong, this is dysfunctional leadership, often times it is dictatorial leadership, and it has never worked long term. The consequences are catastrophic. These types of leaders should change immediately or be removed from their position because NO organization can afford the damage this type of thinking and behaving causes.

I have over 18 years of experience leading and implementing human performance improvement in Energy, Health, Safety, and Environmental Management, focusing on performance improvement initiatives successfully implementing leadership performance improvements through people-centric models, and safety performance improvement in different companies and countries. Through the years, I've observed and experienced the debates regarding the validity, application, and practice of Behavior-Based Safety (BBS). Being a scientist, it seemed only natural to learn about the seminal research, functionality, and true science that drive BBS and to test them and see if they really work. I have discovered that BBS works because it is science; however, I have also found that the safety community has mixed BBS with what I call Results-Based Safety (RBS), and this mixture has confused so many people because they are not able to separate the science from the results-based anecdotal approach.

RBS is steeped in a flawed leadership philosophy that is a top-down approach. Leadership is about influence, but when a person has a title of position but doesn't have the ability to influence, bad things happen. Positional leaders revert to leveraging their positional authority, and they become dictatorial in their leadership style. This is where they practice *telling* people what to do. In my research I found this practice of *telling people what to do* to be the single most significant cause of leaders' dysfunction and ineffectiveness to influence people, but more often than not, they got results! For me, it was a revelation that it's the norm for these types of leaders to be everywhere, they talk at their people instead of with their people, they make people suffer with their endless monologues at safety meetings, make constant emotional appeals, and the worst thing of all—they are so happy with themselves. They feel the struggle is real. Life as a safety leader has purpose and meaning for them. However, if you measure the results, it only works at a great cost. It's "actually" not working. In the safety leader's reality, it seems to be working, but in the actuality of the people and the results, it's not working long term. RBS is short term, and BBS is long term.

BBS gets blamed for RBS, and that's the confusion. RBS is not BBS. Here's a simple way to tell the difference—RBS is about results, and BBS is about behavior. Results are an indicator that you are reinforcing safe behavior: (1) fluently, through expert-level Performance Safety Coaching© and (2) frequently, to achieve habit strength. BBS supports that people are the solution, and RBS sees people as the problem.

My initial realization caused me great concern. The more I asked questions and observed safety practices, it became clearly evident that RBS was the practice and not BBS. BBS is a science, and you need to follow the specific practices specifically. Science is exact. The pushback during my research was unreal, but I soon realized that people at all levels were just frustrated, they cared, and they wanted to improve, and their questions were tough, but they forced me to think and dig deeper to understand what was not working. What was the true root of the problem? It was challenging to sift through so many symptoms, but my research revealed that dysfunctional leadership combined with the practice of RBS is the root problem in the safety industry. This realization regarding the safety industry has shaped my frustration and has driven my research to find solutions, but it also prompted me to write this book to provide clarification on the science of BBS and separate out the dysfunctional practice of RBS.

Everyone thinks they know about BBS, but what they practice is RBS. To make the necessary change in how safety is practiced, you need to understand how human behavior works. For example, if you think you know BBS so well here is a simple question to test your knowledge. How many types of consequences are there to all human behavior? When I ask business and safety leaders this question, they never know, and that's a key indicator they are practicing RBS.

Understanding human behavior will shape how business and safety leaders interact, how safety leaders are selected for their position, how safety leaders shape the safety culture, and how to progressively increase their influence on their team, their people, and their organization. Application of the science of human behavior leads to effective safety leadership and leveraging the incredible performance capabilities of people.

BBS is, however vexed by a dilemma—that people have an amazing ability to adapt to what is expected but can also at times be unpredictable, and at worst behave in ways that are unreliable—yet at its core BBS is about believing in people! In this book, the reader is invited to consider the other side of safety, the pure BBS side of safety that drives safety performance and the effective leadership practices that help organizations utilize the capability of all their people and creates fully supportive cultures that realize high performance. At the core of this approach is compelling scientific evidence that when leaders influence their people in healthy ways and provide reinforcement of desired behaviors, they effectively support and guide real transformation that naturally achieves desired results.

In my engagements I always start with the leadership—their personality, their emotional quotient, their vision and values, and what they believe about their people. Performance improvement always comes through shifting to the other side of safety, practicing pure BBS that focuses on the capabilities, capacities, and potential of people, and aligning performance to the organization's expectations and goals. Many leaders are willing to listen, and when the evidence of the program is presented, they are willing and able to change how they think and act. However, it's not always the case. Some leaders are addicted to their authority and not willing to change to the other side of safety, rethink their role, and do safety differently, but they also do not create sustainable safety performance. It seems that business leaders are okay with this fact. Unfortunately, the employees are forced to suffer while the highly paid leaders continue to enjoy their high quality of life.

Acknowledgments

I'm thankful to a lot of people who played a role in supporting the development of this book. I'm also thankful for the journey itself as a process of learning and learning about so many great people who put their lives on the line every day. I thank my wife, Tania, for her supportive ideas and suggestions throughout the writing process of this book.

Author Biography

Robert Palmer has a PhD in Industrial Organizational Psychology. He is an Organizational Development Leader with 18 years of industry experience. Expert in designing employee experiences based on the science of human performance, performance management, talent management, change management, and organizational design. Dr. Palmer utilizes technology and people science to create organic cultures that help people do their best work by linking performance initiatives to business strategies and outcomes. Dr. Palmer's expertise is in creating agility in the workforce to deliver a specialized customer experience and align to business needs through organizational efficiency, and employee performance. Global project management experience gained from international business projects in India, Israel, and Europe.

Introduction

Safety programs focus on making impressive milestones, for example, a global company in 2021 hit one million hours worked with no OSHA recordable injuries since 2020. The team worked for more than two years (more than 1.2 million hours) without a recordable injury. OSHA defines a recordable injury as an injury that requires more than first aid treatment or results in days lost from work. Safety leadership was proud of the OSHA-defined accomplishment.

The safety leader explained how an achievement like this is a testament to how every single employee stands behind their core value of safety. The safety leader talked about how everyone was empowered as a safety leader and that employees share in the ownership and commitment of seeing that every employee goes home safely every day. However, safety leaders would have a hard time demonstrating a correlation between ownership, commitment, and no OSHA recordable injuries. The real questions that beg to be asked are how did the safety leader achieve those results? How did the safety leader create the safety culture? What safety metrics were tracked?

A typical safety program will have safety leaders say things, like "it's all about putting people first." Safety leadership will shape their talking points around how the safety program has integrated ideas like "inverting the organizational pyramid" and "empowering people in the field." They have "increased their focus on the challenges employees face in doing their jobs each day," which then "results in more meaningful conversations and better feedback which leads to stronger results for the business and our customers." A typical safety leadership team generates concern reports, hazard hunts, and site audits to ensure that safety and quality issues are identified and corrected before an injury occurs. Each month, the employees who submit the most impactful concern reports are recognized and rewarded. The safety leaders talk about how important it is for the people working at the site to "report any concerns or issues they see that could lead to a safety event or a quality miss." The employees' eyes and ears at the site are the best and most important tool they have to preventing incidents.

Safety leaders will talk about their stop work protocol, that "employees have authority to stop work until a problem can be fixed." The Stop Work protocol is encouraged any time someone thinks something doesn't look right or if they don't have the proper tools or equipment to do the job safely. You will hear statements like "we have to go to work thinking about what could happen, and what am I going to do to make sure that everyone goes back home to their family safely at the end of the day, we strive for an accident-free workplace."

Everything just described is RBS, but most people would call it BBS. You have to understand that meaning is derived from context. In safety, context is everything, because it shapes how it is meant to behave. For example, the Stop Work protocol seems like a good idea, and in reality, it gives people the authority to act, but in actuality it's more like giving people permission to avoid an accident. In many ways it's fake safety. Here is the context: how bad was the safety industry at one point in time, to where an individual was not allowed to stop work? Or it wasn't even possible to stop work? Now the great idea is to allow people to stop work—to avoid having an accident, and NOW they have permission. This is RBS at its best because BBS reinforces people to think safely and do what they need to do to be safe. They don't need permission to stop work, and they would stop work!

There are several aspects of the safety industry and safety leadership that need to change in order to be a fully functioning safety program that keeps people safe and productive. First is the lack of understanding regarding the science of behavior. Business and safety leaders talk about behavior, but they don't demonstrate any understanding of behavior. For example, the safety leader whose program achieved over a million hours worked without OSHA recordables said it was because of the ownership and commitment of the employees. However, ownership and commitment are not behaviors, they are outcomes of behavior. How do ownership and commitment get measured? I'm guessing they don't. How does the safety leader define a quality conversation or feedback, and how does the quality get measured?

Second is the lack of clarity regarding behavior and results. The safety leader takes great pride in 1.2 million hours of no OSHA recordables, but that's a result, not behavior. OSHA is a results-based organization that fosters Results-Based Safety programs, and they're not the only ones, this is a major problem in the safety industry on a global scale. The mix of methods in safety that focus on RBS works directly against the science of BBS. Results are not the goal, not the focus, and not the purpose of safety. Results are an indicator. Results indicate that you have defined behaviors accurately, reinforced them fluently and frequently enough to drive habit strength of behaviors, and that the safety culture fully supports safe practices.

Third is the lack of understanding regarding safety culture. Safety culture has to be defined on purpose. Safety culture leadership needs more development regarding how leadership impacts safety culture and how safety culture is shaped. Hiring the right safety leader involves selecting the appropriate competencies, digging down into personality strengths and

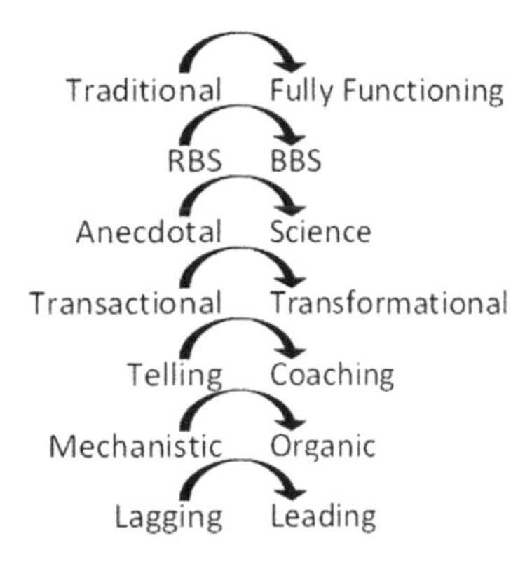

Figure 0.1 The other side of safety

weaknesses that reveal how an individual functions, leadership style, and leadership practices, and understanding the "potential vs capacity" gap to influence themselves, others, and large groups of people. In addition, what is the safety leader's competency level to create a safety culture infrastructure? The safety literature lacks significant research regarding these aspects of safety, but they need to be addressed.

The purpose of this book is to lend perspective, share knowledge, and improve the application to the way safety leaders think and practice safety. Business and safety leaders need to consider the other side of safety—shifting from tradition to a fully functioning model, from RBS to BBS in framework, from anecdotal to science in method, from transactional to transformational in leadership style, from telling to coaching conversations in communication and empathy, from mechanistic to organic in culture, and from lagging to leading indicators in measurement—to see the big picture of how safety is currently happening, take advantage of the insights effectively, and predict and stop all incidents and injuries from occurring (Figure 0.1).

Part I

The problems with the way safety functions

Chapter 1

Ineffective leadership between business leaders and safety leaders

The misalignment between business leaders and safety leaders is natural because each has a different focus. Business leaders tend to leverage their authority and pull rank because safety leaders work for the business leader. This is the root of the problem in safety, that business leaders drive safety in the context of business which is about getting results.

Safety is people-centric and not just results-centric. Results are an indicator that reveals if your safety plan, practices, procedures, and policies are working. When results are the sole focus of the program, the tendency is to ignore the people-centric aspects of the safety program. The safety leader's attitude becomes condescending towards the employees, revealing the safety leader's belief about the employees which is that employees are dumb, employees are to blame, employees don't think, employees make all the mistakes, and it can get caustic to where the culture is dysfunctional. There is also the temptation to manipulate the results to make things look better than they are.

Being people-centric means your safety program's purpose is to support your people to be their best and then check the results to see how you (the leadership) are doing. Result-centric safety programs focus on getting results; the mentality, attitudes, values, and beliefs of the safety leader are completely different from a behavior-based framework. Results-based safety leads to multiple dysfunctional layers in the safety culture. People-centric practices include identifying and planning using the science of human behavior and understanding what motivates people. Safety leaders focus on inspiring their teams and assign significant meaning to tasks that need to get done daily. By being intentional in integrating purpose into the operations, a people-centric safety program enhances a functional culture. Results-centric safety programs reveal significant dysfunctions in safety programs that need to be changed.

DOI: 10.1201/9781003340799-2

THE FOCUS OF BUSINESS LEADERS AND SAFETY LEADERS ARE NOT ALIGNED

Business leaders

The focus of business leaders is business. Business requires aligning talent with their company's organizational purpose, driving performance in all sectors of the business, and ultimately delivering profitability to the board and employees. Business leaders focus on compliance with the federal government and providing a service at a price point that is profitable yet affordable to customers. Business leaders think in terms of bottom-line profitability; sometimes, that is all they think about, and that is not good. The business leader's focus is profits (the quicker, cheaper, and faster—the better), and in general they also have to think about real estate, labor, capital, and entrepreneurship (i.e., innovation) as well as disruptive technologies and competition aspects to keep the organization moving towards the vision. Henry Ford stated, "Business must be run at a profit, or else it will die. But when anyone tries to run a business solely for profit, then also the business must die, for it no longer has a reason for existence."

Business leaders often lose sight of this intrapersonal relationship between profit and purpose. Leadership is the most significant factor in the success of a business. Bad leadership results in a poor performing company, good leadership results in a good performing company, and great leadership results in a great performing company. Everything in business rises and falls with leadership. Great leaders make everything work together seamlessly; without great leadership, all other business resources are less effective. The quality of the leader and the quality of the leadership directly determine how seamlessly everything works together and that's how performance is measured. Regarding safety, the business leader has to be aware of how the science of behavior works and support the people-centric approach in safety.

Safety leaders

The focus of safety leaders is safety. Safety leaders tend to see safety in terms of policy, defined process, process control, solutions, and continuous improvement. They often use methodologies like Quality, Lean, Six Sigma, and Culture of Excellence which are solid business methodologies that seek to find out the root issue of a problem and then provide a solution as quickly as possible to fix that problem. Safety leaders think in terms of the technical. Often, a safety leader is an engineer, many of them have strong military backgrounds, and these factors promote technical thinking which aligns well with business leaders. Business and safety leaders are sometimes similar in that they like to solve problems that are rooted in a defined process, a policy, or a controlled environment.

Often, safety leaders are heavily influenced, to focus on the context of the business leader's perspective of safety-getting results. Business leaders are persuasive in getting their people to focus on the bottom-line results. Safety includes process and policy, but it also needs to be people-centric. You cannot solve people problems using technical problem-solving methods. Business leaders and safety leaders operate from the same business premise—to find problems and then solve problems. Business and safety leaders look at business and the business of safety as: "What is the problem?" "Solve the problem!" "How hard can it be?"

Safety leaders are significantly influenced by business leaders and driven by the fact that they want to avoid being financially penalized with OSHA recordables. This is not wrong in and of itself, but unfortunately, what happens is that safety leaders try to apply business methodologies to human aspects of work, and these methodologies do not work with humans. In fact, they have the opposite effect. Humans are not a business process or a business system. Psychology offers the best insights regarding methods for properly dealing with people in the workplace and specifically safety. This is the expertise of industrial-organizational psychologists and how people think, feel, and act in the workplace.

Safety leadership is more than overseeing the general day-to-day of the organization's safety program. The safety leader needs to be competent in influencing people to perform within the context of the organization's culture and motivating employees to perform their duties and create and sustain the safety culture, which is an incredibly difficult task. They get confused as to why employees cannot be more robotic, just do what they are told, and follow through on what was discussed in the safety training or safety meeting. The frustration is real because the pressure to get results is real. In business the intensity to deliver in a competitive market is a matter of survival. Another dysfunctional aspect of safety is leading with authority.

LEADING WITH AUTHORITY IS DANGEROUS LEADERSHIP

The function of leadership is the most persuasive influence in any organization. A leader is someone who has a title and bears a responsibility to do a thing. Leadership is influence. Leadership is about the ability to influence others to accomplish the goals of the organization.

Leading with authority is dangerous because it is a sign that the leader does not have influence. Without influence, a leader is desperate and has to leverage their authority to control people because they are not able to influence them. Not all leaders have the ability of leadership. Leadership is the ability to cause people to believe something and convince people regarding a situation or an event, by providing a sound reason, for people to pursue a goal.

A leader's role is to provide leadership (influence) through persuading people. When a leader does not have the ability to fully influence through persuasion, the potential for disaster is catastrophic to that leader's capability to be effective in the role of leader, to the people who are under that leader, and to the organization. The leader's ability to influence is based on the power of their authority. Power is a way of rating a leader on how influential a leader is at any given time. A leader's power needs to be aligned with their natural leadership style, their personality, and within the context of the vision, purpose, and values of the organization. A powerful leader is one who has expanded their influence over a large number of people.

According to French and Raven (1959), there are six bases of power: legitimate, coercive, referent, expert, informational, and reward. Legitimate is based on being hired into the role. The coercive basis of power is the one that emerges with an exponential amount of risk because it transfers so easily into authoritative power. A leader who thinks he or she is in a position of authority tends to be driven by ego and arrogance. Authority establishes a faulty premise for all reasoning for action.

The bottom-line profits, financial goals, the market, customers, employees, and the board of directors pull leaders in the direction they want and need. The pressure to get things done and get results motivates a leader to capitalize on every possible opportunity in the marketplace. The pressure of business is what strains a leader's leadership. Referent power is based on respect for the leader, and you identify with this leader's values, personality, and the way they conduct themselves. Expert power is based on expertise and the demonstration of knowledge and ability to act. Informational power is based on having key or significant information. Sometimes leaders bogard information for this purpose; however, informational power can be based on key insights, knowledge, experience, or expertise that no one else has. The risk here is transparency. The lack of transparency is not good. Reward power is when a leader has permission to reward employees with a raise, bonus, or a day off. A reward is about rewarding others and benefiting the hard work of an individual who has achieved their goal or accomplished a result.

A leader does not always have the level of power or depth of leadership (influence) needed for a given situation or circumstance, and this creates complex problems. When a leader lacks leadership (influence), they rely on authority. Authoritative leadership works in some contexts (e.g., military, police), with checks and balances (Schaubroeck et al. 2017). However, authoritative leadership runs the risk of crossing over to an ego driven motivation (Zhang & Xie, 2017). Because of their ego, they become selfish, focused on the letter of the law and not the spirit of the law to control people. Their attitude is condescending, and the rules are not for them, but for the "employees." They believe that they are better than the employees and treat people with disrespect, degrade employees, talk tough with no merit, and implement penalties and punishment to show off their authority. Their sole focus is to achieve a result.

An authoritarian leader struggles to influence people to get things done because people are demotivated and will not follow that leader unless forced. When results are not achieved, the pressure grows to intense levels and threats, blame and accusations are leveraged against workers. Weak leadership creates a vacuum that works against them, and they don't even realize what they are doing to themselves—it's a serious blind spot. This blind spot forces weak leaders to leverage their full authority to dictate policies and procedures, direct the action of people, control all efforts and behaviors, and coerce people with false data and fake narratives because they are desperate and egotistical. It creates a vacuum where performance and participation by employees becomes meaningless and engagement drops into low percentages as evidence that their leadership is not working. Therefore, these types of leaders don't want to participate in employee engagement surveys or other metrics—they're not interested in the truth or in improving because they don't care about the employees, just themselves. This type of leader needs to be fired immediately but is not fired mostly because executive leaders are the same way. It's a leadership silo of ineffective leadership.

The authoritarian leader becomes a bottleneck to their own success and achieving the desired results because their focus is on control, not the vision of the organization. This type of leader is weakened and becomes petty and pathetic. The use of authority works in the opposite of discretionary effort. Authoritarian leaders have the wrong focus. They want full control of the team, and that includes full or partial credit for everything accomplished, and do not tolerate much autonomy within the group. Authoritarian leaders become a barrier to people's performance and their potential. Authoritarians punish and penalize people for failing to comply with their demands, which are often beyond the policies of the organization, and become biased and prejudiced in their behaviors (Rojas Tejada et al., 2011). Authoritarian leaders do not focus on performance. Authoritarians use veiled and unveiled threats and engage in a range of both overt and covert manipulation of employee behaviors, tactics, and even intimidation to try to ensure compliance with their wishes.

The consequences of authoritarian leadership have been shown to be intense stress levels for employees based on factors such as intolerance of mistakes, expectations for complete compliance, and demands for results through expectations that employees should be working harder and faster. Authoritarian leaders consistently raise the bar and abuse power because they are emotionally detached from their people, and they often are unable to empathize with others. Authoritarian leadership fails in the long run, it wears people out, psychologically damages them and causes employees work below performance standards, decrease engagement, or leave the organization in self-preservation. Yet, authoritarian leadership is tolerated, and the authoritarian leader never sees their behavior as wrong, or worse they just don't care.

THE LACK OF DIVERSE PERSONALITY TYPES AND THINKING IS NOT HELPING

Another issue safety programs face is based on the personality types of so many safety leaders. In my work I always have safety people take the MBTI Step II Form Q® personality assessment. Other personality assessments could and should be used as well. Personality assessments are tools that help gain insight into how people function and enhance their capability to reach their potential (optimal functioning). Personality functioning is an area of study that focuses on understanding how individuals become the best that they can be as well as how they may achieve their full potential. Every person has a certain amount of potential to do a thing, and their capacity to do that thing needs to be developed in order to reach their full potential.

In my work most of the safety leaders function as the Sensing Thinking (ST) type personality. I have tested over 500,000 safety leaders. It is an interesting phenomenon that so many of the same personality types are selected for safety leadership. The ST type is a very process-focused individual. They like a linear format for understanding or making sense of their information. Typically, they are incredibly detailed people finding a step-by-step procedure helpful, clarifying, and definitive to move towards a goal. They enjoy creating and following procedures and producing documentation to establish a way of doing things. STs expect others to execute best safety practices fully and willingly. STs are very practical people who desire for things to work easily and smoothly and have a critical sense of subtle differences. Sometimes they tend to see the negative side of things first, but that is not always a negative thing, in and of itself, but a critical differential from the established or defined standard of their normal functioning. It is in their nature to question things and test things, as they want to see the proof that something works. This aspect of their personality, being critical, can become dysfunctional because it becomes judgmental and can be used to manipulate people or a situation to what they want. If STs are not careful, they can create a reputation for only seeing the negative side of things and being excessively tough on people. It doesn't help that they lack empathy for people.

In my experience, ST safety leaders do not like the fact that some people take longer to learn how to adhere to safety behavioral expectations, or they deem some workers are too lazy to do what they've been trained to do, or that people do not need to be reinforced. I often hear them say, "why can't people just do their damn job!" The ST personality type is strong on logic (not necessarily science) and is a routine person; they are often described as robotic or cold in how they behave because they are so consistent. They can, at times, get negative about other people's lack of consistency and usually develop a low tolerance for this "flaw" in other people.

This attitude can negatively impact a safety program by setting a tone in the safety culture that communicates to others by *telling* them what to do. Telling people what to do is a serious hindrance to safety performance and a significant problem. A safety culture that embraces this attitude of *telling* people what to do is detrimental to a healthy safety culture because it drives a results-based focus and not a behavior-based focus. Results-Based Safety (RBS) is the norm in the safety industry, and people continue to confuse it with the other side of safety—Behavior-Based Safety (BBS). Any integration of BBS into an RBS program results in an RBS program. The two programs do not mix. There is science, but you cannot dilute science.

There is a need for more diversity of personality types in safety to help combat the one-dimensional thinking and attitudes that can develop. The iNtuitive Thinking (NT), iNtuitive Feeling (NF), or Sensing Feeling (SF) types could bring a fuller perspective to the table and a broader capacity to think beyond a linear process. These other personality types would also find great value in considering how safety policy, procedure, and practice affect people. The NT, NF, and SF personality types can bring new innovative approaches to safety, and they also can effect positive changes to a safety culture that focus on optimal functioning capacity. The lack of diverse personality types fosters group thinking, and group thinking leads to the silo affect, which are strong indicators of RBS.

These other personality types bring a more collaborative experience to the team. Diverse personality types create diverse team functions, and that supports the agility to quickly form new teams and build more meaningful, inclusive relationships once everyone's functioning is understood and applied appropriately. The RBS culture misses the people-centric perspective—this is a serious problem that safety leaders are lacking awareness about. People feel like they are being treated like robots, and they do not like it. You cannot treat people like robots or assets and view everything through a lens of policy and procedure, which is a tendency of the ST personality type, and sustain successful outcomes.

Chapter 2

Practicing results-based safety

GOOD INTENTIONS ARE NOT SCIENCE

The safety industry is properly intentioned in that they want to stop all incidents and injuries from happening again. The intention of safety leaders and their programs is to figure out what went wrong and fix it, fix the specific thing that went wrong, and make sure it does not happen again. This is the traditional side of safety that started with the industrial revolution in 1750–1760 and the invention of the steam engine. Employees and/or the public would get hurt, and engineers would seek to solve the problem and prevent it from happening again.

However, employees continue to be injured at high rates for our day and age. The Bureau of Labor Statistics (2020) reports that there were nearly *3 million workplace injuries* (many life-altering and life-ending) in 2019 alone. Eisenbrey (2013) states that these incidents cost U.S. businesses 250 billion dollars a year. The Bureau of Labor Statistics (2020) shows that although workplace incident rates have steadily declined by 28% over the last decade, rates for serious injuries and fatalities (SIFs) *have remained virtually unchanged.*

The traditional side of safety is confused. Traditional safety leaders are governed by their business leaders and are pushed to get results so the business leader can look successful and not cost the company money, but they are also pulled by the safety community to get results—hence the confusion. Safety results are an indicator that the safety culture has properly defined safe behavior(s), aligned to desired safe behaviors, and that are providing the necessary reinforcement with frequency and fluency. That should be the focus, but it is often not the focus. Safety leaders become obsessively focused on safety numbers (results), but not the behaviors of themselves or their people. These results are measured in two buckets of indicators: leading indicators and lagging indicators. Results-Based Safety (RBS) is measured in lagging indicators, but Behavior-Based Safety (BBS) is measured in leading indicators. The other side of safety is about embracing and integrating the science of behavior into safety, measuring using leading indicators, and reviewing results as an indicator that the appropriate

DOI: 10.1201/9781003340799-3

amount of behavioral reinforcement is frequent and fluent enough that the desired safe behaviors are achieving habit strength.

The first glaring problem with Results-Based Safety is that the incident or injury has already occurred. So, this is a reactive effort. The second glaring problem is that searching for a root cause is a systematic or process approach when it should be a behavior approach. The confusion between results and behavior science happens because the safety culture is designed for results, and not safe behavior. In most safety cultures things get rolling once an injury or incident has occurred. The investigation focuses on finding the root cause. At this point in time, finding the root cause is not a bad thing; however, it is a sign that the science of behavior has not been utilized properly from the beginning to prevent incidents and injuries from occurring. The other side of safety is the behavioral science side, and it seeks to prevent and predict incidents and injuries from occurring based on the science of human behavior.

Over the years, the traditional side of safety has created a body of knowledge with intricate methodologies, procedures, processes, and policies to keep people safe. However, because the traditional side of safety focuses on "what happened?" it can be frustrating trying to understand the root cause because it is not based on a methodology, procedure, process, or policy, but on behavior. The frustration mounts as business leaders bear down on safety leaders to get things fixed and get results. The pressures of corporate and federal politics, business expense budgets, insurance costs, legal proceedings, potential bad press, federal fines, and organizational (business leadership) embarrassment cause people to react, to pursue a result. Another exasperating problem is that safety policies and procedures often set safety leaders up to chase symptoms of the root cause, and not the actual root cause. This phenomenon happens often because of time constraints and pressure to get results.

The traditional side of safety is all about pursuing results, but this cannot be the only focus, and this is where everything goes wrong. The other side of safety, the behavioral science side is about behavior. This is the significant difference between the two sides, and it is the most misunderstood aspect in the safety industry. You must understand that BBS looks at results as an indicator (i.e., a leading indicator) that the correct number of healthy reinforcements is occurring and that the number of desired and defined safe behaviors is being observed.

Pursuing results, in the traditional sense, leads to reaction and opens the door to all the pressures of the safety industry and business leaders to get it fixed. The other side of safety is about implementing behavioral science and is responsive to pursue a predictive solution to incidents and injuries. Pursuing results causes a myriad of negative and unhealthy outcomes including the ultimate worst practice in all of safety: manipulating results. The focus naturally becomes chasing symptoms rather than pursuing the root cause because of business demands (e.g., it costs too much money and takes too much time). If a root cause is discovered, the focus is singular, a singular incident, a singular result, and that is hardly the big picture. Because the goal is not behavioral (i.e., improve desired safe behavior(s) that drive safety outcomes), the focus is to get a result (RBS). When your goal is to achieve a result and you fail, it causes frustration. When business and safety leaders are frustrated, they leverage their authority and lash out at people and blame people. There are numerous negative outcomes that surface as indicators of an unhealthy safety culture—this is not BBS, this is RBS. Knowing the difference is crucial to developing fully functioning people, and a fully functioning safety program.

With enough pressure even the calmest of safety leaders becomes frantic and frustrated, and blaming people is natural, and it is a sign of practicing RBS. The people are the problem, and now the focus morphs into a pursuit to get people to comply with rules, policies, and procedures (this is the opposite effect of BBS). Safety leaders cite regulations and demand *letter of the law* compliance to regulations to get their results. BBS gets blamed, "it does not work," "BBS is about blaming people," and "BBS is bad," but this is not true because you are actually practicing RBS. Safety leaders who do not understand BBS and do not correctly implement behavioral protocols are the ones failing. A program does not fail, but leaders fail. The execution of a safety program is the leadership's responsibility.

Blaming BBS is a sign that you are utilizing RBS practices. BBS is about behavior. When RBS is being practiced, safety leaders become weak leaders. Weak leaders take on an authoritarian tone, make threats, demand respect for their authority, and express most of their communications with a punitive tone in their voice. Weak leaders develop and express a condescending attitude, and instead of asking people questions and seeking to collaborate with them, they *tell* people what to do, and everybody gets blamed, even those who are practicing safety and did nothing wrong. Therefore, workers do not trust the safety leader. It sends a message that "I am better than you," us vs. them mentality. I, the great leader of safety, did not make the mistake, but you did! It communicates that the leader needs to achieve the safety result so that they can earn their bonus. By the way, the amount of trust is an indicator of whether you are building and sustaining healthy relationships or not. Trust is an outcome. So many people mistake trust as a commodity just like safety results, but trust is an outcome that indicates you are in a healthy relationship.

Some basic comparisons of BBS vs. RBS are presented in Table 2.1.

Table 2.1 BBS and RBS basic comparisons

BBS	*RBS*
Clearly defined behaviors	Clearly defined results
Coach people	Blame people
Practice reinforcement	Practice punishment
P^r N^r R^e	P^u P^e P^x
Empathize with people to think safely	Tell people what to do and be done with it

USING LAGGING INDICATORS IS THE WRONG PREMISE FOR PROBLEM-SOLVING

The traditional safety way of doing things starts with the wrong metrics, using lagging indicators only. Business leaders pressure safety leaders to get results, and safety leaders want to keep their jobs and get paid their bonus money, so they pursue results. This focus on results skews the true essence of Behavior-Based Safety (BBS) by not focusing on behaviors but focusing on safety results.

Business and safety leaders together have unwittingly shaped habitual behavioral routines that pursue results in their safety programs. What is being practiced is a culture of Results-Based Safety (RBS), and this is the major problem with current safety programs—the all-out pursuit of results. The pursuit of safety results is not effective and does not work or you would not have the level of politics that takes place in safety or the lack of transparency.

Results are an indicator; they are needed to let you know if the safety program is on track. Are you doing the right things? Are you trending correctly? Are you achieving the goals that are required of the program? Do you even have the right goals? However, you cannot create a safety result by trying to manipulate that safety result. Looking at lagging indicators has good intentions, but it is obvious that the incident has already occurred. Once an incident, near miss, or injury has occurred, it is too late to prevent it. The safety goal, bonus money, or job of the safety leaders and safety of people are at risk, and therefore the pressure to get results is ratcheted up with urgency and dominates as the motivator for safety. Stop calling out the failures of BBS when what is really being practiced is RBS.

Business and safety leaders must address the other side of safety. It is time for them to look at and focus on the other side of safety—shifting from tradition to basics of a fully functioning model, from anecdotal to science in method, from RBS to BBS in framework, from transactional to transformational in leadership style, from telling to coaching conversations, from mechanistic to organic in culture, and from lagging to leading indicators to see the big picture of how safety is currently happening. Take advantage of the insights of the leading indicator safety data, and seek to effectively predict, and stop all incidents and injuries from occurring.

Using lagging indicators only does not result in successfully preventing incidents and injuries from occurring. This is not a logical or a scientific approach if you understand the science of human behavior. Unfortunately, the natural outcome of trying to stop bad things from happening lends itself to focusing on the negative. Catching people doing unsafe acts, recording those unsafe acts and near misses, utilizing lagging indicators to make improvements, and, worse, punishing and/or penalizing people for committing unsafe acts do not lead to improvement of safe behavior. Using a punitive to get more from people is scientifically backward. Punitives stop the behavior. You can't stop bad behaviors and assume that safe behaviors will take their place. Safe behaviors have to be defined and reinforced until they achieve habit strength. Business and safety leaders are actually behaving in a way that gets them the opposite of their desired results, and the unintended consequences are a punitive culture where: (1) people are not reinforced to *think* about behaving safely for themselves but are reinforced to comply to policies and procedures, (2) people hide their unsafe behavior, so they will not be caught, and (3) relationships are rarely healthy because people are not treated respectfully, which results in a lack of trust or no trust at all between safety leaders and employees.

Many safety leaders make the mistake of thinking a cordial relationship is somehow a healthy relationship. I have observed thousands of safety interactions and 99% of them are awkward. Cordial but awkward because the trust is not there. Trust is an outcome of healthy interactions (a relationship is the result) and does not look or feel like most people think. Trust is an indicator of a healthy relationship. All this effort is being made all around the world, and safety leaders are leading programs that work against their very own desired results.

It is not intentional, but most safety leaders and their programs are confused as to what they really support and the goals they are intending to achieve. As well-meaning as safety programs try to be, they encounter so much unscientific advice and that has set up safety leaders, entire safety programs, and entire safety cultures to fail. The outcome of all this noise has had a significant number of unintended outcomes. For example, you have probably heard of the Zero Today safety campaign that many organizations adapt. The intention is to have zero incidents or injuries; however, the very title of the program is a giveaway that the focus is on a result—zero incidents or injuries. This is unscientific. This whole campaign misses the point of Behavior-Based Safety (BBS) because it is Results-Based Safety (RBS).

PURSUING RESULTS IS RESULTS-BASED SAFETY

The focus on results is the worst safety practice of all time. It encourages the worst behavior in leaders, and nothing good comes from poor leadership. Anyone can get the safety results they want; it is easy to achieve, but it is often unethical, immoral, and in some cases against the law, and it is not scientific. Business leaders are often the ones who push, even bully safety leaders into pursuing RBS because in many cases all they care about is getting those numbers and getting their bonus (which is a result-based incentive). Executive leaders need to think about the impact their incentives have on business leader behavior, but then again how often have you heard,

"I don't care how you do it, just get me my numbers." Business and safety leaders who pursue RBS demonstrate several behaviors that indicate that results are their only focus:

1). Leadership spends more time coming up with creative safety numbers that meet their goals than they do the actual practice of safety. Rarely do business leaders celebrate what safety leaders do when they have a positive impact on their employees or have conversations about how they reinforce safety thinking and behavior. Traditionally, the conversation between the business leader and safety leader is focused on how to minimize the impact of a recordable incident, and how to communicate it (cover) and who to blame for this situation.

    If the safety program makes it a few weeks or months without any incidents or injuries, it is like being on pins and needles trying to make it to the end of the year without another recordable incident, until BOOM! There is a recordable incident that happens. Then the blaming starts and everyone will be stressed out. The Results-Based Safety mentality is all about making the numbers. It produces a blaming culture. Business and safety leaders seek to creatively massage the safety numbers with minimal metrics. Ironically, their sole focus on lagging indicators is what hurts them. I have been in many meetings where business leaders argue against the validity of appropriate statistical analysis of safety science, and it is painfully obvious they do not fully understand statistics, psychometrics, or anything about human behavioral science. Business or safety leaders not willing to listen, not willing to significantly understand how human dynamics play a role in safety are in fact putting people's lives at risk.
2). The safety campaign is their major safety effort. The safety campaign is an antecedent. It prompts or reminds people what to do, but it does not cause a behavior to happen. The campaign usually includes a workshop or planning meetings that bring all the safety leaders together. The campaign's focus at these workshops is a big emotional pull on the heartstrings of the safety people. Business and safety leaders share tragic stories of death that make you feel sad, even depressed! When I listen to these leaders share these anecdotal correlations of how safety leaders need to do MORE, and everyone needs to do MORE so that lives can be saved! It is impossible to achieve any success based on doing MORE.

    It is rather pathetic to think you can scare, motivate, and appeal to people as if dying is something they do not care enough about. I would not want an anecdotal doctor performing surgery on me. I also would not want an anecdotal business or safety leader leading the safety efforts that involve me. It is true that safety is serious business, but doesn't that mean that these leaders need to spend the time and money to understand the science of human behavior in safety? Should not every effort to learn, hire the appropriate people, purchase the software, apps, equipment, personal protective equipment, and everything else be part of the solution to all these tragic deaths? Applying the science of human behavior is critical to safe behavior, striving for best behavior practices, and achieving safety success should be the focus, not a safety workshop that's just part of a safety campaign that ultimately undermines Behavior-Based Safety.

It costs lots of money to produce and implement big, colorful, creative safety campaigns, and it makes business leaders and safety leaders feel like they are doing a good thing, and there is nothing wrong with a safety campaign per se. However, a safety campaign, a safety workshop, or a safety training, for that matter, is an antecedent. Antecedents prompt or remind one to do a behavior, but it does not cause a behavior to occur, and therefore these efforts have little to no impact on required safe behavior. If you can't statistically correlate the impact of a safety workshop or safety campaign with improved safety results, that's not good.

It is not wrong to have a safety campaign, but it needs to be designed in a way that is aligned with required safe behavior and supports a measurable behavioral outcome. How is it that a business or safety leader does not know this? RBS and BBS are not the same, and it's critical that the difference is understood. While lives are on the line, how tragic is that ignorance continues at the highest levels of safety leadership.

3). Business leaders and safety leaders act out of anger and punish people (publicly and privately) for not achieving results. These types of practices combined with emotional outbursts and embarrassing displays of anger and frustration demonstrate that one is not the right fit for the job. Regular Outbursts of anger are a demonstration of low emotional intelligence. Business and safety leaders get caught up in blaming individuals for ruining their results. I have seen business leaders turn over tables, pound their fists on desks, and scream at a single individual in front of an entire group. Regular Emotional outbursts are unprofessional and cause immeasurable psychological damage to the people who strive to be safe every day, and need psychological safety (Edmondson, 1999). It diminishes safety performance, but since there is no measurement, there is no proof of the fact that business and safety leaders have diminished safety performance. What happens, most often, is that the data (lagging indicators) show these emotional outbursts seemingly improve safety results. Emotional leaders can get people to perform in the short term, but if you review their history, you will see a pattern of up and down roller coaster safety results (Edmonton). If an analysis of their safety program were allowed, it would look horrible and pathetic for them, and transparency is buried deep under layers of politics and red tape.

The emotional outbursts seem to work because people are steeped in fear. They are scared to make a mistake, scared to lose their job. A majority of safety leaders think this is a good thing—that people work in fear, but it is not a good thing. It makes you wonder why employees do not quit their jobs for fear of dying. Usually, because these RBS types of leaders pay their employees more money than they deserve. It is a bribery system. Getting paid to be abused is a trap, and the business or safety leader is reinforced to continue having emotional outbursts because it helps them achieve the safety result, they want. It's a sick daily cycle that causes physical, emotional, and mental damage to safety workers.

4). The focus of the safety program or campaign is on the short term and not the long term. A short-term focus is another clue that the safety program is focused on results. Business leaders often do not genuinely care about sustaining the safety program because in many cases, they hope they will not be around long enough for a disaster to strike. To get quick wins, a business leader will offer incentives. This is a non-scientific practice because rewards are linked to results—RBS. BBS is all about reinforcing behavior both positively and negatively, and a sustained safety culture requires a scientifically based approach that focuses on safety as a lifestyle—BBS. In a BBS program, results are an indicator that you have defined the desired and required safe behaviors accurately, you reinforce those behaviors at the proper frequency, your safety people fluently perform Performance Safety Coaching© conversations, the safety training learning objectives are aligned with the goals of the safety program, and your leading indicators cover the broad spectrum of all aspects of safety. The safety of people should truly be about reinforcing the habit of thinking safely and acting safely, no matter where you are or what you are doing. Always thinking about being safe and acting safely.

Business leaders and safety leaders do not intentionally mean to cause confusion, but it is not acceptable to be well-meaning and follow unscientific advice and/or methods. There is a desperate need to implement the science of behavior and not rehearse tragic stories of lost loved ones. When safety leaders comply with business leaders and pursue only results, it causes a tremendous amount of confusion regarding safety in general. Confusion is bad leadership.

RBS is misleading. A true BBS program seeks to implement psychological principles of behavior to influence the way people think, feel, and act and to reinforce every individual to think safely and act safely about the situations they will be and are in now. It is about awareness to plan and think safely no matter where you are, not just at work. However, RBS is so powerful because in the end business and safety leaders make big money if they achieve results—which are powerful motivators. The effect of RBS is subtle, yet pervasive. RBS creates a false reality, and it makes business and safety leaders think they are accurately addressing the safety issues in their organizations, but it is all in the context of results. For example, one safety campaign slogan stated *be safe at work, home, and play.* This campaign has good intentions, but it is still about results and is not focused on safe behavior. Safe thinking needs to be a way of life. Safe behavior is the result of reinforcement that achieves habit strength. Safety results are the outcome of safe behavior.

5). The emphasis of policy and procedure is over the value of people and quality relationships. Another aspect of RBS vs. BBS is to consider how safety leaders like to use policy and procedure to hold people accountable for their actions. Having well-defined policies, clear step-by-step procedures, mapped best practices, and visually stimulating safety models are excellent practices that establish what is required of people to be safe. The problem is that many safety programs are not clear on what is required or expected.

Business and safety leaders confuse people with their policy and procedure when they act contrary to that documentation or to the values espoused by the organization. For example, when one of the corporate values states *we treat each other with respect*, but the safety leaders treat people with disrespect-people get confused. People tend to resist policy and procedure that are forced on them because they are being forced, and now they do not trust management or leadership. How could anyone trust the leadership that treats them with disrespect by blaming them for poor behavior that the system supports and reinforces? Leaders are responsible for the systematic safety that is in place within their organization. The objectives of the safety program need to be in alignment with desired and required safe behaviors, consequences, goals, best practices, reporting, metrics, and results, all in a way that people can trust one another.

The culture of the safety program is a significant indicator if business and safety leaders are getting it right. This is the worst outcome that safety leaders can achieve in the world of safety—an unhealthy safety culture because unsafe behaviors will continue to occur, without anyone figuring out why. This continues to be a big mystery in safety. The fact that no matter how much money is spent, committees created, policies written, and procedures established, people continue to get hurt and safety leaders do not know why. So? They blame the individual. It must be the individual. Safety leaders think they have done everything they can do. What else could there be?

6). Behavior-Based Safety (BBS) is focused on defined desired and required safe behavior(s) that are reinforced (positively and negatively) until they achieve habit strength. That is a big missing aspect of most safety programs—they do not identify the desired and required safe behaviors but focus on what is NOT desired (e.g., do not do this, do not do that). A review of a company's safety policies, procedures, training materials, videos, and safety records will reveal a consistent use of negative terminology. More importantly, you will rarely find any reinforcing terminology, reinforcing metrics, or the use of leading indicators.

The unintended consequence of these efforts is a dysfunctional safety culture that feeds into an outcome of zero trust and does not achieve zero incidents or injuries. An RBS culture blames the individual, and the focus is on finding someone who must take responsibility and be punished—even fired! It is a good excuse for weak leaders to feel like they are in control and doing something good. A Results-Based Culture feeds a business leader's or safety leader's ego, and they see themselves as above everyone else, and it is not their fault and therefore somebody has to pay.

If the business leader and the safety leader are not willingly playing an active role in safety efforts, then they see themselves as above everyone else. Unfortunately, if they have any kind of safety incident, it gets covered up. Because, the whole system is flawed, and it is not a person's fault, but everyone's fault, and the whole system needs to change. The system will not change if safety leaders keep practicing the same old methods and asking the same old questions.

Part of the problem with safety programs is that safety leaders continue to ask the same old question, "what went wrong?" The problem with this question is fourfold: first, if this is all you ever ask, "what went wrong," there is a bigger problem, which is that the overall comprehensive safety system is not designed to stop incidents or injuries, but it's designed to fix things in order to get results. Fixing things is not getting to the root issue of the problem. Too often, asking what went wrong leads to chasing faux issues and phantom symptoms that are not the root problem. For too many years safety leaders have struggled to understand and implement behavior-based practices because of the myths, misconceptions, and well-meaning but unscientific advice of business leaders, safety leaders, safety people, and safety consultants.

Second, safety leaders start with a lagging indicator(s) and work backward based on "what went wrong." The fact that the safety event has occurred means safety leadership is now working in the past, and it is impossible to prevent that specific event from occurring once it has already occurred.

Third, when the event is recorded and saved in a database, it usually includes a detailed description of the injury, not the behavior. Rarely is the data collected ever about the behavior, why would you not want to collect that data? Why would you not be interested in that data? This type of situation presents an opportunity for more diverse thinking on so many levels.

If the safety leaders are, for example, all engineers, then you only have an engineer's point of view. An Industrial-Organizational psychologist could be extremely helpful in establishing behavioral measurements. The lagging indicator data collected is seemingly supportive, but it is only a partial picture of what is really happening and ends up only supporting Results-Based Safety. The whole point of Behavior-Based Safety is to focus on desired and required safe behavior(s).

A fourth issue is the unrealistic expectations of business leaders. When you focus on results, you demonstrate that you have no understanding whatsoever, as to how difficult it is to work with human behavior. Business leaders ignorantly demand results from their safety leaders and that puts undue pressure on them to produce numbers. This is not the point of any safety program! It is the desired outcome, but it is not the focus of the program. This is a serious misunderstanding in safety. Business and safety leaders need to understand that results are a manifestation of the fact you are reinforcing the correct behaviors in the correct way (i.e., frequently and fluently) and measuring identified and defined leading indicators correctly.

Most safety leaders never focus on behavior because they do not understand how behavior works. The world of safety needs to start seriously comprehending and engaging in the other side of safety. The other side of safety is the leading indicator side. A behavioral scientist should be involved. Just like having engineers do engineering work, lawyers doing legal work, and industrial-organizational psychologists should be doing psychological work. Behaviors(s) should be included with all incident and injury reports because you need to track behavior(s).

Safety leaders and safety programs focus heavily on lagging indicators only. What safety leaders do not understand is how to define a behavior and how to record a behavior, and ultimately how to track and trend behavioral data. An industrial-organizational psychologist would be extremely helpful and know how to record and track behaviors. What is frustrating is that safety leaders only focus on lagging indicators based on events that have occurred in past tense, and now their data on safety events are lagging and only indicate what has gone wrong in the past. The logic, at this point, that is implemented is that incidents and injuries can be mitigated going forward, because safety leaders, after investigating, now know what not to do, and therefore plan, strategize, and campaign on what not to do in order to be safe. If you think about the logic of that logic, it does not add up. It is an impossible task to stop things from happening by trying to stop things from happening. There are natural forces at play, and science provides an insight that we can measure, and therefore trust.

The singular focus on lagging indicators creates an endless cycle of trying to fix what went wrong. The intent is good because safety leaders and safety programs are seeking to prevent negative events from occurring. The endless cycle goes on and on because it is difficult to fix what might go wrong based on what did go wrong. The pursuit of lagging indicators has become embedded in the global safety culture. Safety leaders have become so accustomed to this cycle that it has become almost impossible to change their minds that there is another way. Ask any safety leader what metrics they track, and the answer is almost always lagging indicators.

The problem with lagging indicators is that it is too late to prevent the event. Lagging indicators only demonstrate that something has gone wrong. What is so frustrating is that safety leaders desire to fix what has gone wrong, but they are chasing symptoms and not the true root cause of the problem. Their intentions are good, but it is impossible to get to the root of the cause of the safety event because the root cause does not exist in the lagging indicators. The root cause exists in the behaviors of people.

For example, a current standard safety practice is to correctly fill out a tailboard. However, it is also a common practice that incomplete or inaccurate tailboards (e.g., pencil whipping) continue to exist. Safety leaders get frustrated, and rightly so, but they focus on fixing this problem by *telling* their people to "get it right," or "fix it." Either way, they are missing the point that motivates the required safe behavior in this example. There are numerous moments every day to reinforce and motivate people to behave to the expectation, but you have to reinforce that behavior.

What results from *telling* people what to do is the continuation of inaccurate and inconsistent tailboards. To "fix the problem," the safety leader resorts to "coaching" (which is incorrect because it's actually "telling") and they coach (*tell*) their people in an accusatory or blaming tone trying to change the result—a correct and accurate tailboard instead of an inaccurate tailboard. I guess coaching doesn't work! What is not understood is that filling out the tailboard accurately is the behavior. Safety leaders need to learn how to coach (converse) with the individual about the performance of a person properly, regarding the desired or required behavior, not *tell* them what to do.

Safety leaders must learn how to reinforce behavior positively or negatively depending on the attitude of the performer, and their performance towards the desired or required safe behavior. Reinforcement has to occur frequently, and the coaching has to be fluent in the conversation for every individual so that they learn how to do the safe behavior, over and over again, and over again until it becomes habit strength. That is the key to improving safety results, by reinforcing the safe behavior until it is at habit strength. If you are doing performance coaching right, you will be on the right path to getting the right results. Results are an indicator that will indicate if you are reinforcing safe behaviors correctly, appropriately, accurately, and achieving habit strength of those behaviors.

Safety metrics rarely include behavioral reinforcement metrics. Result-Based Safety (RBS) focuses on results, and Behavior-Based Safety (BBS) focuses on behavior. Behavioral reinforcement data must be collected and measured to learn and understand the trends of safe behaviors. Positive reinforcement (P^r), negative reinforcement (N^r), Recovery (R^e), Punishment (P^u), Penalty (P^e), and Extinction (P^x) would be leading indicator metrics, and they need to be tracked. With the safety people I've had the opportunity to work with, I have been able to define around 168 leading indicators.

For example, in my BBS 5.0 Safety Program, I teach safety leaders to observe a worker or group of workers in action, and regardless of what is observed, whether it is desired or non-desired behavior, you would engage in a Performance Safety Coaching© conversation via the seven steps and review the desired or required safe behavior with those you just observed. Every coaching observation is an opportunity to reinforce the proper safe behavior. The individual or group would receive a positive or negative reinforcement depending on the behavior that was observed. Statistically speaking there exists a correlation between reinforcing (P^r/N^r) the desired safe behavior, and the frequency of the desired safe behavior occurring. The more frequent a desired safe behavior occurs, the fewer the incidents or injuries that occur. The goal is to reinforce the desired or required behavior until it reaches habit strength. A habit is a strong behavior and exceedingly difficult to change, regardless of if it is a good or bad behavior. Imagine a workforce of people working in a culture of reinforced desired or required behaviors that are all at habit strength. That is the ultimate vision of a great safety program.

Unfortunately, most safety programs do not collect any data on these specific efforts of behavioral reinforcements. Nor do they practice Performance Safety Coaching© conversations. A major reason is that safety leaders have incorporated Results-Based Safety, and not Behavior-Based Safety. RBS focuses on results, and BBS focuses on behavior.

I once consulted with a safety leadership team responsible for the safety of over 15,000 people. I reviewed five years' worth of safety data, and there were no records of any kind regarding behavioral reinforcements, no data regarding coaching conversations, and no leading indicators whatsoever. They strictly focused on lagging indicators. However, the safety leadership team was adamant that they were a Behavior-Based Safety program. They were not! And there was little to no evidence of BBS. They were a Results-Based Safety program. That was incredible to me that they could make such a claim, and not have any defined safe behaviors, zero recorded reinforcements, or zero data from coaching conversations. Now I know what they mean by Zero Today! Zero is a result. How can you say you have a Behavior-Based Safety program if you are not monitoring, tracking, or measuring any behaviors? This makes no sense to me, but I see it happening everywhere in the safety industry. Business and safety leaders are confused, but so are safety consultants and psychologists. They bash away at BBS, but it is actually the practice of Results-Based Safety, and they don't get what's happening in their safety culture.

What they did have at this company was a long history of reoccurring safety events—including the death of employees and a strong history of pursuing results. The lagging indicators were revealing that the safety leadership was not getting it right. You need lagging indicators, but more importantly you need leading indicators.

My research has focused on defining desired and required safe behaviors and increasing the number of observations of desired and required safe behaviors and increasing the number of reinforcements (P^r/N^r) for those identified and defined safe behaviors.

The focus on safety also includes *fluent* and *frequent* Performance Safety Coaching© conversations (fluency and frequency are leading indicators) and creating the infrastructure that measures leading indicators. What results is a focus on desired and required safety behaviors, an increase in those desired safe behaviors, and a decrease in the number of unsafe behaviors and events. The correlation is significant.

It is time for safety leaders to consider another way, to consider the other side of safety. It is time to look at and focus on the other side of safety-shifting from tradition to a fully functioning model, from anecdotal to science in method, from RBS to BBS in framework, from transactional to transformational in leadership style, from telling to coaching conversations, from mechanistic to organic in culture, and from lagging to leading indicators to see the big picture of how safety is currently happening, and take advantage of the insights to effectively predict and stop all incidents and injuries from occurring. There are other intangible benefits as well, such as an improved safety culture with an increase in Performance Safety Coaching© skills, improved respectful tone of voice in conversations, improved levels of empathy, improved openness to receive feedback, improved quality of relationships, increased employee engagement, increased trust in leadership, increased discretionary effort towards safety in general, and increased engagement in thinking, planning, and being safe.

However, there are inherent risks to this approach. The biggest risk is the involvement of business leaders. If business leaders only have an interest in results, then the path to superior safety performance is very difficult if not impossible. Business leaders must change their relationship with the safety leader by understanding Behavior-Based Safety leadership. What does that relationship look and sound like? Link those safety leadership behaviors to competencies and reinforce safety leadership behaviors. Together, business and safety leaders can develop and support a safety culture to accomplish the following: (1) comprehending the science of human behavior at a high level, (2) aligning safety behaviors with competencies, (3) linking required expectations and behavioral outcomes, (4) setting and communicating the expectations of the program, (5) allowing all designated safety people to attend training, (6) support enforcement of accountability at all levels and not take the attitude that Performance Safety Coaching© and tracking reinforcements are for field people only, and not office people, (7) purchase technology (e.g., an app to collect the data), and of all things, and (8) the willingness of everyone in safety to buy into and correctly practice the science of behavioral reinforcement, which is the true essence of Behavior-Based Safety.

The use of Artificial Intelligence, Robotics Processing Automation, data mining, machine learning, digitalization of routine tasks, and mobile apps is making it possible to collect, study, and trend behavioral data like never before. Creating a safety dashboard is essential to becoming predictive in safety trending. Predictive analytics is an actual science that uses models to extrapolate patterns in historical or transactional data. This is helpful in identifying risks and opportunities to improve in the behavior areas that may not be receiving enough reinforcement. It is important that relationships in the data are captured and assessed for risk or opportunity. The relationships of the data can provide keen insight that behavioral conditions exist in the environment that guide decision-making regarding high and low risk strategies, policies, procedures, focus areas, reinforcements, consequences, antecedents, and potential outcomes.

Chapter 3

Creating a false reality

In the traditional pursuit of safety, to get results, business and safety leaders have mistakenly developed a false narrative in their safety culture. They talk in terms of safe behaviors, but they practice Results-Based Safety (RBS). The gap between these two methods is significant to the impact and influence of a safety program. For example, the RBS methodology has leaders who fight against the protocols in the science of human behavior—creating a false reality that their safety program is working, but in actuality it is not.

A false reality is a tricky, subtle, and complex layered thinking entity. First, understand that *reality* is the state of things as you perceive them. Reality is based on your idealistic thinking. Reality is notional—it's influenced by your personal theory, self-suggestions, and the invention of mental frameworks and fantasy. Reality is the sum or aggregate of one's perceptions that are existent within the subjective context of the individual, and the system of work. Safety culture is a *contextual* system. Context involves the circumstances that form the setting for an event, statement, or idea, and in terms of which it can be fully understood and assessed. A system is a set of things working together as parts of a mechanism or an interconnecting network. A contextual system, like safety, needs to consider the relationship between individuals and their physical, cognitive, and social worlds, but they rarely do that because the problem with *reality* is that it can be created or based on the ontological status of your thinking, and thus seem "real" when they are not real or not actual. For example, a vast majority of safety programs are premised on a reality that is not based on how things are, but rather on how they are made out to be or thought to be. The reality is a façade and not actual.

Actuality is based on the actual existence of facts, data, figures, numbers, and science as contrasted with what was intended, expected, or believed to be in existence of a thing. Actuality is objective, is meaningful, and measures significance. Actuality is realized when you apply the science of behavior to the safety culture context, and behavior *actually* changes.

Actuality requires a different data set, not just lagging indicators, but a shift to the other side of safety (e.g., understanding human behavior, motivation, consequences, performance coaching, leadership, and leading

DOI: 10.1201/9781003340799-4

indicators) all based on actuality. Most safety programs do not collect data related to the other side of safety, and in actuality the safety culture is based on a false reality. The false reality is heavily reinforced, and therefore it seems so real to the leadership and the employees that the majority will fiercely defend it. It is interesting to note that "reinforcement" is being utilized, but safety leaders do not *realize* it, and they are not aware they are actually reinforcing the wrong things (e.g., wrong attitude, emotions, thinking, and behavior). It is difficult to explain to safety and business leaders because for years they have only understood their reality, and the actuality is a completely different concept.

The move to the other side of safety is a crucial one, a necessity. This is a critical gap for businesses, and the burden of responsibility is on the shoulders of safety and business leaders to understand the gap between reality and actuality, but also how reality and actuality need to be aligned to create an authentic safety culture. How do you make this crucial shift? You create a proper infrastructure based on behavior and competency that if done correctly is revealed in its outcomes (metrics) proving safety success.

Business leaders, safety leaders, consultants, and even psychologists have developed some strong negative attitudes about the science of behavior in the industry of safety. In the end, people get hurt because these fake experts do not know science. It is the same thing when people talk with authority about medical conditions, procedures, medicinal ingredients, or prescriptions. You may have read about it, but you are not an authority, you are not a medical doctor, and therefore you are potentially dangerous to people who listen to someone like you. Certain personality types have discovered a shortcut to feeling important, and that's speaking with authority. It gives them instant credibility and feeds their ego, but they are dangerous and need to be checked.

This is often the case when it comes to Behavior Based Safety (BBS). Fake scientists or even some psychologists describe it as blame the worker, carrot-and-stick approach, it is about catching people doing something wrong, and on and on it goes. There are 54 divisions or types of psychologists all defined by the American Psychological Association, and very few are qualified to work in safety. Why is it that non-scientists, fake scientists, business leaders, and safety leaders think that behavior is easy? Why does the functioning of humans seem so simplistic to them? And yet, when you try to have a conversation with them, they get authoritative, loud, and even angry about the topic. They are quick to reference their 30, 40, 50, or 100 years of experience in safety. They know what is best, just follow them and do what they tell you to do and get those results. RBS is what they are practicing, not BBS, and it's detrimental to the safety industry.

Here is an example of fake safety guidelines from a large insurance company representative presenting at a conference about how to create an effective safety program.

Presentation: 10 steps to create an effective safety program

1. *Ensure management's commitment to safety*
   What does this even mean? On a scale of 1–10, what would a manager's commitment be? Has anyone in management ever responded publicly, "no," they do not support safety? Even if they say "yes" and give you money for your colored posters and fancy campaign, does this improve safety results? This is a meaningless suggestion because it does not define or measure anything regarding the commitment of management to safety. Management's commitment to safety could be measured and used as a leading indicator.
2. *Designate responsibility for safety*
   This is great! Should it be one special person? How about a group of special people? How about a diverse group of special people? How about everybody! Everybody is responsible for safety. Does this improve safety results? This is a meaningless suggestion. However, if you could define responsibility and define the behavior(s) that demonstrate responsibility, these could be used as leading indicators.
3. *Determine the safety requirements for your workplace*
   How does one do this? What does this look like? Well, let me suggest to you that it looks and feels like a lot of rules. Most safety leaders have a negative mentality—they see the glass half empty. They have learned that being negative (critical and cynical) has created an image where people see them as a leader. A critical spirit is not good leadership and usually contributes to a negative safety culture, and this is rarely addressed by business leaders towards safety leaders. Does making lots of safety rules improve safety results? No, not really, and this is a meaningless suggestion, and unfortunately it allows negative people to be in charge and is a practice of RBS. What should be practiced is BBS by clearly defining what desired safety requirements look like? What does a good safe day look like? This could be a leading indicator and should be measured.
4. *Conduct a hazard assessment*
   If you are in safety and do not know what this is, you are pathetic and will probably die today. If you are in safety and do not conduct hazard assessments, and this is a new idea to you, you will probably die today. The suggestion here makes it sound like conducting a hazard assessment is a one-time activity. Conducting a hazard assessment should be a leading indicator and should happen on a regular basis, as a consistent behavior. Does conducting hazard assessments improve safety results? Do you have a clear definition of a hazard assessment? Do you have any metrics that capture the number of hazard assessments? Do you have a metric to capture what a proper (based on definition) hazard assessment looks like? You should. This is a metric that should be measured as a leading indicator.

5. *Develop a written safety policy*

   This is a ridiculous suggestion. Legally you must have a written safety policy. How would a safety leader not know this? Does this written safety policy improve safety results? It should if you write it correctly. By the way, lawyers are not industrial-organizational psychologists, they need help in writing a safety policy, and it needs to include behaviors.
6. *Ensure two-way communication about safety*

   This is a basic and lame suggestion. Does two-way communication improve safety results? Most safety leaders are negative, and if two-way communication is a baseline, well then, dame it! Three-way communication is better! Hell, four-way communication is even better! Wow! Such impressive leadership! Do you have any metrics that demonstrate two-way, three-way, or four-way communication improves safety results? You should. This is an outcome that should be measured as a leading indicator.
7. *Correctly identified hazards*

   Seriously? This is obvious—no? How sad, but I have literally seen with my own eyes safety programs that identify hazards with no plan, procedure, or policy that clearly defines how to correct the identified hazards. Do you measure the number of identified hazards? Do you track the number of corrected hazards? Do you record the percentage of identified hazards to corrected hazards? Does correcting identified hazards improve safety results? This is an outcome that should be measured as a leading indicator.
8. *Train employees in safety*

   This is a very benign statement. I wonder how many meetings were required to get this suggestion put on the list. By the way, do you track how many people you train? Do you track test scores? Do you have test scores? Do you collect feedback on the learning objectives? Are your learning objectives behaviors? Do you test how well people have learned the material? Do you measure pre-knowledge and post-knowledge scores? Do you measure training impact? Why not? The training itself is an antecedent, but the outcomes should be measured as leading indicators.
9. *Keep workplace hazard-free*

   Yes, agreed, but how do you define hazard-free? How do you define metrics for hazard-free? Would keeping the workplace hazard-free help people and the environment be safer? This is a metric that should be measured as a leading indicator.
10. *Review your safety program and keep up to date*

    What a bland suggestion. However, how many times do you review your safety program—once, twice a year? What about conducting a SWOT (Strengths, Weaknesses, Opportunities, and Threats) Analysis on your program? What about creating a general survey about your safety program that the employees take? What about a leadership survey about your safety program? These are outcomes that should be measured as leading indicators in your safety program.

These ten steps are actually Results-Based Safety and do not create an effective safety program and, in many ways, are fake safety. You experience this all the time at conferences, safety training, webinars, and in the workplace. What is key to understanding is that these ten steps are anecdotal and not science. Anecdotal means that it is not necessarily true or reliable, because the insights are based on personal accounts rather than facts or research. The very definition of anecdotal is that it is not science. Anecdotal efforts usually end up as a "pulling on heart strings" safety campaign. These are emotionally based campaigns that are not science. Safety campaigns should be behavior campaigns. How do you measure these ten steps? How do you conduct measurements that notify you that safety is occurring every moment of every day? Most safety programs only track lagging indicators.

One could argue that not getting the science right is safety negligence. When Results-Based Safety is practiced, there is negligence regarding the science of behavior in safety. Negligence is the failure to take proper care in doing something. Negligence is a failure to exercise appropriate and/or ethical ruled care expected to be exercised among specified circumstances. There are five elements of negligence: duty, breach of duty, cause, in fact, proximate cause, and harm. Business leaders must support safety leaders in being confident that there is no negligence in their safety programs. Leading indicators contribute to the minimization of negligence.

Part 2

Applying the science of behavior

Chapter 4

What motivates behavior?

Psychology is the scientific study of behavior, cognition, and emotion. Behavioral science, which is based on how psychology looks at the subject of human actions and explores the cognitive processes and behavioral interactions between people in their natural work environment. It involves systematic analysis, investigation of human behavior through naturalistic observation, and controlled scientific experimentation. It attempts to accomplish legitimate, objective conclusions through rigorous formulations and observation.

You practice science or you practice being anecdotal. A majority of safety programs are based on an anecdotal premise. Being anecdotal is not scientifically true or reliable but is based on personal accounts rather than facts or research. Business leaders and safety leaders love being anecdotal because it is all about them, their experience, know-how, insights, and favorite stories, and it helps feed their larger-than-life personalities, but it also makes the work environment less safe.

The age-old question: why do people do what they do?, is still worth asking and understanding. Especially in safety, understanding the why could save countless lives and protect people from injuries. Science is the tool that helps us understand what motivates people to do what they do and seeks to understand why. Motivation is a key aspect in the world of safety, and it is important that safety leaders understand motivation, and how to motivate their people to act safely.

Motivation is the proclivity to initiate, continue, or terminate certain behavior(s) at a particular time and is strongly influenced by organizational culture. Motivation is a state of being which is subjective to forces acting within and upon the person that creates a disposition to engage in goal-directed behavior. Safety leaders need to understand the role that the safety culture plays in influencing the motivation levels of their people, and how individuals are motivated. There are four types of motivation:

- Extrinsic (external source, action)
- Identified (external source, non-action)
- Intrinsic (internal source, action)
- Introjected (internal source, non-action)

DOI: 10.1201/9781003340799-6

EXTRINSIC MOTIVATION

Extrinsic is defined as not being part of the essential nature of someone or something, coming or operating from outside. Extrinsic motivation then is focused on what is outside the person. Extrinsic motivation has four major components: power, popularity, pleasure, and perfection. Because extrinsic motivators are extrinsic, are easily observable, and so are their defense mechanisms which are used often for avoidance and protection of the major components such as repression, regression, denial, reaction, projection, rationalization, or displacement.

External regulation is extrinsic and refers to motivation that is regulated by external influence via intimidation, penalty, punishment, compliance, or conformity. For example, an employee who practices a safe behavior exhibits external regulation because he or she is instructed to by their safety leader. Identified regulation which involves awarding a conscious value to a behavior in such a way that the action is accepted when it is personally important (Ryan and Deci, 2000). This means that the safety culture could be experienced in such a way that being told to act safe requires that a person accept that "*telling*" or command immediately and change their behavior to comply with the "*tell*" or command. External regulation could be the cause of fake safety because an employee only behaves safe when the leader is present and telling the employee what to do in the immediate moment makes them fake their compliance. This is a dangerous cultural element to many safety cultures, and it needs to be addressed.

IDENTIFIED MOTIVATION

Identified motivation refers to a form of motivation, which occurs as understanding or feeling the need to perform or accomplish a task, but not yet acting on this need. This is a powerful form of motivation as it is intrinsic to the person and prepares the person for acting when enough motivation builds to a threshold level and action occurs. Often, motivation is a building process where need and desire are actualized. For example, if an employee could identify the consequences of a risky behavior, they could be motivated not to take that risk. The need and desire to choose to be safe often takes time to actualize depending on the context of the situation because of fight or flight syndrome—fight is often based on courage (response), while flight is often based on fear (reaction). This form of motivation is powerful because when actualized it potentially creates lasting accomplishment or performance enhancement. However, it is also a risky delicate balance. The risk is courage vs. fear. Do you want your people to react in fear or do you want your people to respond in courage (i.e., confidence, awareness, being present, and actively thinking)? This level of motivation is often impractical to leaders because they don't want to wait for this process to develop in individuals. A safety leader needs to understand the motivational development of their people. Motivation is not instantaneous every time like most people think.

INTRINSIC MOTIVATION

Intrinsic means to belong naturally; it is essential to the nature or constitution of a thing. Intrinsic motivation refers to an internal motivation that is natural and, more importantly, essential to the individual. This form of motivation is subjectively based on the individual and is believed to occur when behavior aligns with values, and enjoyment is experienced when doing a task. This form of motivation can be triggered in a variety of ways. One example would be providing reinforcement that reflects the value of the individual. The key to understanding this internalized motivation is to align a person's vision, values, beliefs, and desires in relation to the goals of the individual, and to define the performance requirements so the individual knows the minimum of success.

INTROJECTED MOTIVATION

Introjected means the unconscious adoption of ideas or attitudes of others. Introjected motivation is an internalized motivation like intrinsic motivation, but it is based on non-action. Introjected motivation is how we passively learn from others, their good or their bad habits. It transfers to use via observation, and we are motivated to change our own behavior in order to model the behavior of others. The underpinning premise is based on positive reinforcement (P^r), negative reinforcement (N^r), penalty (P^e) or punishment (P^u) depending on the individual. The stimulus for the person's motivation is internal rather than external. This form of motivation is more common than people might believe and takes many forms such as leaders criticizing in a negative and/or sarcastic tone of voice the poor job someone performed, blaming people, putting people down, and making statements that are intended to induce feelings of guilt within people to motivate them to perform better. This form of motivation has many negative aspects as it can anger or confuse people when the person experiences constant negative interactions or by not being able to satisfy themselves via their behavior or the person causing them to feel negative never stops making them feel negative.

Motivation plays a key role in safety culture, employee engagement, and individual, team, and organizational performance. These forms of motivation provide insightful perspectives on how people function, but more importantly, they provide different ways to access the motivation of the self and others. Safety leaders should be well trained in the ability to motivate people properly for personal and organizational improvement.

EMOTIONAL INTELLIGENCE

Another aspect of motivation that is lacking in the safety industry is the use of emotional intelligence (EI). The measure of one's EI is done by using an emotional quotient assessment. Emotional intelligence is the ability to monitor one's own and others' feelings and emotions, discriminate among them, and use this information to guide one's thinking and actions (Salovey & Mayer, 1990). One's EI is an excellent way to understand how an individual functions emotionally, learn areas of strength and reasons for gaps in performance, and comprehend potential areas for development in how one experiences the work culture, and how they operate when working with others.

Safety leadership needs to understand how emotional intelligence can be utilized as a condition that effectively shapes the safety culture. People have to be emotionally motivated to perform at their best. It's one thing to understand one's emotions, but it's another thing to control them, and it's yet another thing to control them and deploy them effectively towards a goal, purpose, and a great vision. The practice of safety needs to include emotional intelligence in how the safety leader develops their leadership style and that style conditions the safety culture in how communications, policies, procedures, and everyday practices of safety are conducted. Emotional intelligence is a competency of safety leadership. Emotional intelligence supports the capacity and capability of a safety leader's ability to expand their influence.

One of the most significant applications of emotional intelligence is the understanding and effective application of empathy. Empathy is the ability to acknowledge the feelings and emotions of other people. Often, empathy is confused with an apology. When hearing some bad news, people will state, "I'm sorry," for some reason, but that's not empathy, that's an apology. Empathy is a link between us and others and is a significant factor of influence to the safety leader because it is how an individual leader can understand what others are experiencing as if we were feeling it ourselves. Proper empathy is an ability to acknowledge the feelings of others. Empathy is when you acknowledge what you see in someone's face, or when you acknowledge what you hear in someone's voice. For example, when someone looks stressed, you can express empathy by stating, "your eyes looked focused, and you looked stressed, do you want to talk about it?" "You look overwhelmed, how can I help you?" Or it could be something you hear in someone's voice. For example, "you sound frustrated, I'm willing to listen if you think it will help." "You sound upset, talk to me." Empathy leads to practical action, and therefore an exercise in intelligence. Empathy is not just an expression but a form of communication that observes and recognizes an emotion and expresses care regarding that emotion that seeks to help by being ready to change from a negative emotion to a positive emotion via a potential solution. This is what drives a bond between people. This is the serious work of leadership—keeping people focused on the positive. Leadership seeks to influence people to focus on the positive, strive for the goal by honoring the values, and complete the vision. Authoritative leadership is ignorant of empathy because the focus is results. Authoritative leadership is ignorant of emotional intelligence because it lacks intelligence. However, the use of empathy opens the door to an understanding regarding the use of consequences. Consequences should be used in the context of empathy.

Chapter 5

Focusing on the utilization of consequences

Safety leaders must focus on creating a culture, through performance coaching, that supports and reinforces people to think through required and desired safe behavior. Establishing the proper cultural conditions is a competency that is missing from leadership everywhere. A safety culture needs to focus on the proper utilization of six types of consequences: three are reinforcement ($P^r/N^r/R^e$), and three are punitive ($P^u/P^e/P^x$).

Safety programs must utilize the science of human behavior via the ABC model of human behavior in order to create a healthy culture (Figure 5.1). The safety culture should demonstrate several health indicators. For example, quality relationships should be established and enhanced by the safety culture. This can be accomplished by the practice of asking open-ended questions and not closed-ended questions (conversing rather than telling people what to do), and therefore the quality of the interactions/conversations that are happening continuously improves by (1) the *frequency* of reinforcement of desired and required safe behaviors, (2) the *fluency* of the Performance Safety Coach© utilizing the seven steps of the performance coaching conversation, and (3) identifying and measuring all these interactions as leading indicators to create data trends and learn what the impact of these coaching conversations are having on the individual employees, to learn if these performance coaching conversations are focused on the right behaviors and which behavioral areas are at risk and are the desired and required behaviors being addressed in a way that is effective and is the data showing proof of sustained safety outcomes.

The ABC model of behavior is based on the work of Thorndike and Skinner and is defined: A is antecedent, B is behavior, and C is consequence, and a behavioral event usually happens in that order. First the A, then the B, and then the C.

It all started with Pavlov (1897) who established classical conditioning where he discovered that associations between events can be learned. However, Thorndike (1898, 1905) took the knowledge further and established operant conditioning where an association is made between a behavior and a consequence and called it the *Law of Effect*. His theory emerged

DOI: 10.1201/9781003340799-7

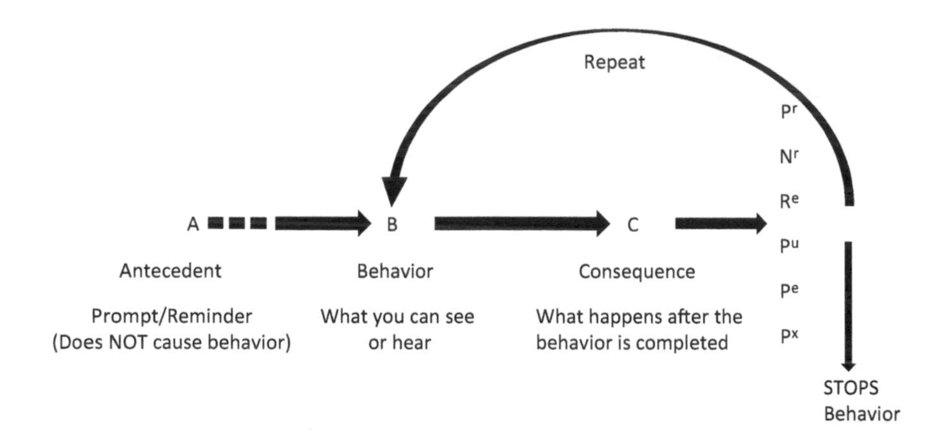

Figure 5.1 ABC model of behavior

from his research and further established that responses that are immediately followed by a satisfactory outcome become more strongly associated with the situation and are therefore more likely to occur again in the future.

Skinner (1938), based on the work of Thorndike (1898, 1905), further defined operant conditioning as a method of learning that occurs through an association between a particular behavior and a consequence. Reinforcement emerged from Skinner's research as an enhancement to the Law of Effect. Skinner established that when a behavior is positively or negatively reinforced, it tends to be repeated or strengthened, and a behavior that is not reinforced, but rather treated with a punitive consequence (i.e., punishment or penalty), tends to be extinguished or weakened.

Schedules of reinforcement are an important component of the operant learning process. When and how often you apply a reinforcement has a significant impact on the strength and rate of the frequency of the response to the reinforcement. The strength of response is measured by accuracy, duration, frequency, and persistence. There are two types of reinforcement schedules: continuous and intermittent. With intermittent, there are four possibilities: fixed ratio, variable ratio, fixed interval, and variable interval.

Another aspect of consequences to consider is that primary reinforcers often occur naturally (e.g., food, water, sleep, air, sex, and safety). Safety is a primary reinforcer both physically and psychologically. People naturally want to be physically safe, but they also want to be psychologically safe (Maslow, 1943). Think about how Behavior-Based Safety (BBS) really works. First there is the physiological needs, second the safety needs both physical and psychological, the third need is belonging, the fourth being self-esteem, and the fifth and final level is self-actualization. Meeting these needs are signs of a healthy safety culture and could be used as leading indicators.

Secondary reinforcers include the verbal and visual cues that have been the main focus of our discussion. A healthy BBS program seeks to apply primary reinforcers through these five levels of needs and then builds upon those needs with secondary reinforcers. The goal is to help people to be fully functioning in the desired safe behaviors.

Results-Based Safety (RBS) is what ruins the primary hierarchy of needs because it is based on results, and the secondary reinforcers rarely get attended to because the primary needs are never met, and therefore there is no foundation to build upon to reinforce desired safe behaviors.

In addition to the ABC model of behavior, there is further evidence of the power of reinforcement based on the Premack principle. Developed by psychologist David Premack (1959), it provides critical insight into human behavior. Understanding and utilizing this principle allows you to arrange contingencies that motivate less desired behavior. The Premack principle states that a person will perform a less preferred activity (low probability behavior) to gain access to a more preferred activity (high probability behavior). A less preferred activity is defined as one in which the individual is unlikely to choose to do on their own, thus developing the term low probability behavior. A more preferred activity is an activity that the individual would likely choose to engage in on their own, a high probability behavior.

When a high probability behavior (high-P) is made contingent on the engagement in a low probability behavior (low-P), the high probability behavior serves as the reinforcer for the low probability behavior, making that behavior more likely to occur. This is usually presented in safety training as a first/then statement or visual (first ______, then ______). The key to utilizing this principle effectively is to ensure that the high-p activity is actually high-p at the moment it's presented to the learner. Preferences for activities change frequently for different reasons so it's important to be aware of the learner's current motivation for activities.

The Premack principle's simple nature makes it a great choice for safety leaders to implement by using the following fidelity checklist:

- Identify the low-p safety behavior or activity
- Identify the high-p safety behavior or activity
- Present the contingency to the learner either verbally or with visuals
- Wait for the learner to complete the low-p safe behavior or activity (continue to withhold access if the learner fails to complete the safe behavior or activity)
- Grant access to the high-p safe behavior or activity once the low-p safe behavior is completed

The steps are straightforward, but it's important to observe the behavioral process for fidelity and to ensure accurate implementation (use the fidelity checklist above). Treatment fidelity is a measure of the reliability of the administration of an intervention in a treatment study. It is an important aspect of the validity of a research study, and it has implications for the ultimate implementation of evidence-supported interventions in typical clinical settings.

The greatest challenge often comes with accurately identifying a behavior or activity that motivates the learner for the "then" part of the contingency. Alternatively, safety trainers and leaders need to provide accurate visuals with this intervention, or they inaccurately use the visual as a schedule and fail to ensure the high-p/low-p sequence.

Understanding a learner's motivation at any given moment is the single most effective way to utilize the Premack principle effectively. The Premack principle requires motivation for a specific activity to make it a high-p activity. Ensuring motivation for an activity necessitates an understanding of factors that influence reinforcer effectiveness including motivating operations (MOs) (see Chapter 4).

MOs alter the current effectiveness of an item or activity as a reinforcer (or punisher) at any given moment. They also alter the current frequency of behavior that has encountered reinforcement (or punishment) with that item or activity in the past. Motivating operations have two effects on behavior: (1) value-altering effects—change the value of a specific consequence as a reinforcer or punisher and (2) behavior-altering effects—change the current frequency of behavior that has been reinforced or punished in the past. There are two types of MOs:

- Establishing operations (EOs)
- Abolishing operations (AOs)

Essentially, as they relate to the Premack principle (Premack, 1959), MOs are factors that influence the effectiveness of an item or activity as a reinforcer and may evoke behaviors that have produced the item or activity as a reinforcer in the past. Establishing operations makes an item or activity more effective as a reinforcer and evokes behaviors that have previously been reinforced by that item or activity. Abolishing operations make an item or activity less effective as a reinforcer and abate behaviors that have previously been reinforced by that item or activity.

There are limitless possibilities available when using the Premack principle. Each one must be specific to the learner's motivation at any given moment. Below are some examples of the Premack principle in safety:

- First safety meeting, then tailboard
- Get your tailboard done, then you can work

The Premack principle is considered an antecedent intervention because it reduces the impact of common antecedents on behavior. In other words, presenting a contingency that utilizes the Premack principle makes maladaptive behavior less likely to occur. This is the nature of an antecedent intervention. Premack explains further regarding response deprivation. It refers to a model for predicting whether access to one behavior (the contingent behavior) will function as reinforcement for another behavior (the instrumental response) based on the relative baseline rates at which each behavior occurs and whether access to the contingent behavior represents a restriction compared to the baseline level of agreement (Cooper, Heron, & Heward, 2019).

Presumably, restricting access to a higher preferred behavior (the contingent behavior) will act as a form of deprivation, serving as an establishing operation. This consequently makes the opportunity to engage in the higher probability behavior an effective arrangement of reinforcement for the lower probability behavior. Response restriction is the key factor in determining whether access to the contingent response would be reinforcing; the reinforcement effect will occur only when the condition of response deprivation is present in the contingency (Konarski, Johnson, Crowell, & Whitman, 1981).

The practice of BBS requires an understanding of the science of behavior, which is a learning process in which new behaviors are acquired and modified through their association with consequences. When safety leaders title their safety programs, they need to be accurate and true to the science, and many safety programs are not BBS, but RBS or some other form of anecdotal premonition altogether. To add to the confusion, the endless use of buzz words instead of science and the focus on a result is semantic satiation, which is a psychological phenomenon (i.e., it occurs in your safety culture) in which repetition of a word or phrase loses its meaning for the listener, who perceives the word or phrase as meaningless.

The practice of RBS has no meaning to people. Business and safety leaders have forgotten that people do not care about what you know or want until they know you care about them. At the heart of safety is a belief mechanism based on "what is meaningful," that you have a dedication, and a commitment to the application of safety science that impacts all the people who do the work. Numbers rarely motivate people, but doing something meaningful always does. People are moved to act, motivated, when something is meaningful to them. Meaningfulness is a consequence and a great motivator. Practicing reinforcements is a direct demonstration of caring (i.e., believing in people and empathizing with people) because it is supportive of the desired safe behavior.

MAXIMIZING THE IMPACT OF REINFORCEMENTS

Positive reinforcement (P^r)

A positive reinforcement is experienced when you receive a desirable outcome. When you click a button (behavior) and it works (consequence) is why you would push the button again. Because the outcome is what you want, you have a strong tendency to repeat that experience and the behavior gets repeated. If that behavior is repeated enough times, it becomes a habit. For example, if you're wearing your hard hat (behavior) and a heavy wrench falls and hits you on the head and you experienced no injury (consequence), you will have a strong motivation (extrinsic) to continue to wear (repeated behavior) your hard hat.

Negative reinforcement (N^r)

A negative reinforcement is experienced when you want to avoid something because it is negative to you, and by completing the desired behavior, the negative thing goes away. You need to get up in the morning (behavior), and the alarm clock rings (antecedent) to wake you up, the ringing is annoying because it is loud and obnoxious (N^r) until you hit the snooze button (P^r), and the ringing (antecedent) starts again annoying you (N^r) until you get up. Once you are up, you turn off the alarm. The negative reinforcement (N^r) of the alarm reinforced you to complete the behavior of getting up. Hitting the snooze button is a positive reinforcement (P^r) to you because you get a few more minutes to lay there, but it is a delay to the desired behavior of getting up. The key to understanding the negative reinforcer (N^r) is that the desired behavior occurs in conjunction with a negative consequence (a loud and obnoxious alarm that keeps going off, until you get up) and therefore is categorized as a reinforcer. This is not the same as a punitive because the very definition of a punitive is to get a behavior to stop. A reinforcer gets a behavior to repeat.

Recovery (R^e)

Recovery occurs when you are learning a new behavior and because there is not enough positive (P^r) or negative reinforcement (N^r) occurring, the new behavior is not supported with enough reinforcement, and therefore the old behavior is recovered because it is the stronger habit. When you're trying to improve your health by exercising more regularly, you set your goal to walk three times a week. You walk a mile each day, then you walk two miles each day. You increase your walking to two miles daily. One day you feel extremely tired, and you don't walk. It was nice having a break from the walking. You walk again the next day, but you're sore so you skip another day. Three days go by, and you feel better, so you walk again the next day but only one mile. You skip a day, and the next day you walk a mile. After a week, you only walk three times that week, and then the next week you walk one day. You don't walk anymore, it's really not your thing. The new behavior, walking, has gone extinct. You didn't receive enough positive or negative reinforcement.

MINIMIZING THE IMPACT OF PUNITIVES

Punishment (P^u)

Punishment is experienced when you receive something you do not want and therefore the behavior stops. At work, a man forgot to wear his Personal Protective Equipment (PPE), and you are forced to wear a bright colored pink safety vest that states, "I forgot to wear my PPE." This is potentially a punishment (P^u) for you because you are made to stand out and/or be made an example for your failure, even be embarrassed for not performing as expected, and you realize that you need to stop forgetting to wear your PPE. Keep in mind, if this does not bother you, then it is not a punishment. Punishment only works if the experience is significantly negative to the individual. This is often why management will continue to increase the punishment in some way (e.g., make you wear the pink vest at the front gate, so everyone can see you), because it does not seem to be working, and they keep increasing the punishment until it hurts enough to get you to stop what you are doing (e.g., forgetting to wear your PPE).

The application of a punishment needs to be thought through. There is a time and place for punishment, but too often it is the *go-to move* by business and safety leaders. Safety leaders end up hurting people by making them feel (e.g., embarrassed, dumb, worthless, incompetent) less of who they are. In actuality, a safety leader, using punishment or penalty, is focused on the wrong thing—typically bad results, but is at the same time reinforcing a culture that makes people feel less and perform at a lower level than what they are capable of. Business and safety leaders must create conditions in their culture that build competencies (i.e., use reinforcement) not reduce competencies (i.e., use punishment as their go-to move). Make people feel (e.g., stronger, smarter, sharper, quicker) more of who they are capable of being—being their best and striving for full functioning.

Penalty (P^e)

Penalty is experienced when something you want and/or value is taken away from you. If you forgot to wear your PPE to the job, and you are dismissed from that project or sent home, you are being penalized because the opportunity to work (to make money) is taken away from you. The desired outcome is to get you to stop doing what is undesired and behave in a desired way. What most safety leaders fail to do is to practice reinforcement. You have to reinforce the desired or required behavior, not just penalize it. Again, there is a time and place for penalty, but it should not be the "go-to" move of safety leaders.

Safety leaders who do not make this shift from penalizing the undesired behavior, to reinforcing the desired or required behavior, demonstrate that they do not understand the consequences of what they are doing to their own people and its impact on the culture of the entire business unit. They leverage punitives, but safety leaders must understand the difference between a punishment (P^u) and a penalty (P^e) and how and when to use each and then shift to the reinforcement of the desired behavior. Reinforcing the desired behavior needs to be the focus. Sometimes the path of correcting a behavior can start with a punitive, but then it needs to shift to a negative reinforcement (N^r) and then shift again to positive reinforcement (P^r). The progression builds trust, it demonstrates that you care about helping the individual improve their behavior and performance. It's not about a result or a number, it's about people. And if you treat people right, they will improve. They will deliver, and the numbers will indicate you are doing the right thing—reinforcing your people to be fully functioning.

Extinction (E^x)

Extinction is experienced when a behavior receives no P^r or N^r and over time the behavior slowly disappears into extinction. For example, in a safety meeting, the safety leader asks people to speak up about safety ideas. You bring up safety ideas, and the safety leader starts to poke fun at you about being the safety idea guy … "Mr. Innovation," and none of your ideas are ever implemented. Over time you stop suggesting safety ideas. Bringing up safety ideas has gone extinct. What's worse is that the safety leader doesn't even realize why you don't bring up safety ideas anymore.

The use of punitives works in the short term, but they are not an effective way to create or sustain a business or safety culture. Too often punitives are the only consequences business and safety leaders use. There is a time and place for the use of punitives (P^u, P^e, and P^x). It is a significant thing to understand that punitives have a negative impact on people, and punitives require precise alignment with how the individual functions (i.e., their personality) or they do not work. A punitive has to be meaningful to the individual to work. You cannot just randomly use any punitive or broadly use a punitive on everybody. The use of a punitive has to be specific to a group, or an individual. It is frustrating to everyone when you practice punitives as a first resort and not a last resort. Punitives affect an individual's motivations, and therefore it must be understood how to use them to be truly effective when practiced in a limited capacity. The problem is that using punitives is reinforcing to business and safety leaders. It makes them feel in charge, and that's the wrong feeling to have. It's not about you as a leader—it's about your people learning to think safe all the time and being fully functional in their role.

The proper application of consequences is part of the conditions that shape the safety culture. The proper application of consequences aligns well with science. A healthy safety culture fulfills Maslow's hierarchy of needs; safety leadership should be focused on creating conditions in the safety culture that meet (1) the physiological needs of the workers, (2) the physical and psychological safety needs of the workers, (3) the belonging (and inclusivity) needs of the workers, (4) the self-esteem (social reinforcement) needs of the workers, and drive towards the (5) self-actualization (self-reinforcement) needs of the workers.

Very few safety leaders, business leaders, consultants, and even a lot of psychologists truly understand this science enough to implement and maintain it correctly. This is the main reason BBS is ridiculed because it is misunderstood at a scientific level and therefore it is applied improperly, and even worse it is mixed with RBS. People attack BBS because of the RBS factors but do not realize that they are actually attacking RBS. They do not know the difference and therefore think they are the same, but they are not. BBS is about behavior, and RBS is about results. You have to know the difference, or you do not truly understand BBS.

In safety programs across the world, BBS is rarely implemented with the proper scientific depth, the rigor of the ABC model of behavior, accurate training that teaches how to coach to precise behavioral expectations, application of proper reinforcement through performance coaching techniques, and precisely defined leading indicators.

One thing that would help people practice the science of behavior better is the clarification of terms. There is a lot of confusion around terminology regarding BBS. It is a significant giveaway that BBS is not properly utilized when the terminology is not used correctly. Let's clarify the terminology.

CLARIFICATION OF TERMS

Antecedent: a simple antecedent is a prompt or reminder of a behavior. It is important to note that a simple antecedent does not cause a behavior to happen. The antecedent is strictly a prompt or reminder.

Behavior: something that can be observed. A behavior can be seen or heard. Often, BBS programs and other safety programs do little or nothing to define behaviors or to define them accurately.

Consequences: a result or effect of an action or condition. There are consequences to behavior(s), and to culture. A true BBS safety program takes into consideration the impact of consequences. It is crucial to understand the interaction of behavior and consequence. There are six consequences known to all humanity. A large majority of safety and business leaders do not know what they are, how sad. The six types of consequences need to be implemented at a cognitive, social, and visual level as a condition in the safety culture. Consequences are a science-based complexity that requires keen insight into the scientific methods required for implementation, a strong accuracy to the scientific principles that enable successful practices, and an ability to implement in the context of cultural development/fit while leading to performance at desired levels. Here are the six consequences.

Consequences are categorized into two distinct groups: reinforcements and punitives. There are three types of reinforcements, and they are called reinforcements because they get a behavior to repeat. When a behavior repeats, it is categorized as a reinforced behavior. There are also three types of punitives. A punitive gets a behavior to stop. If a behavior stops, it is categorized as a punitive behavior. We know a behavior was reinforced if that behavior is repeated, and we know a behavior experienced a punitive if that behavior stopped. People often confuse the idea that good behavior is reinforced, and bad behavior experiences a punitive, but that is not the case. If a bad behavior is reinforced, it will repeat. If a good behavior experiences a punitive, it will stop. Consequences are based on the individual's experience and therefore every experience and every consequence must be looked at independently.

Reinforcement: is a physical action or verbal expression shaped by tone, attitude, mood, or energy level that is directed towards a desired behavior (what is seen or heard). You can only reinforce a behavior. You cannot reward a behavior. A reward is based on a result, not a behavior. This is a huge misconception between a behavior and a result. With a reward, you are rewarding an outcome (result), not a behavior. The difference sometimes comes down to timing, but the distinction is clear—reinforce behavior and reward result.

Encouragement: the action of giving someone support, confidence, or hope. It can involve persuasion to do or to continue something. It is not reinforcement because it is not focused on a specific behavior.

Reward: something earned when a result is achieved. A behavior has a consequence. The subtly here is overtly clandestine because a reward and a reinforcement can occur at the same time. A result is based on the outcome of multiple behaviors. If you want a specific result, then you must define the criterion or standard for the behavior, reinforce that behavior (i.e., positively, or negatively, and be careful about recovery), and when the behaviors are completed, it achieves the desired result. Once that criteria or standard is met, the result is achieved and therefore a reward is administered. The subtle difference is that a behavior by itself does not always yield a result. You have to reinforce a behavior, and then a chain of behaviors in order to achieve the result. Behavior needs to be reinforced. If you want the result to be sustained, then you have to reinforce those behaviors until they are habit strength. A reward encourages you to perform, strive for that result again, or even do better than the last time, but the performance is actually based on the specifically defined behaviors. This is why athletes practice a "move" (behavior) over and over again, and musicians practice arpeggios over and over again, because the perfected behavior produces a high level of performance. Performance gets rewarded, but it's the behaviors that make the performance possible.

Recognition: is based on a result. Special recognition is deserved for those who make special efforts and achieve a unique accomplishment or level of performance. Recognition is also based on the fact that an individual or a team achieved a performance result over a period of time—they are recognized for their achievement. It is a mistake to think that one can reward a behavior. Technically, you reward the result (not the behavior) with a reward, and if the result is sustained over a period of time, they are recognized for it (i.e., achieving the result, not the behavior). The confusion between the two terms makes for inaccurate science and makes it difficult to measure. You need to be clear on what reinforcement actually is and what a reward actually is.

Celebration: is focused on achievement that happens over long periods of time or is based on the significant effort that results in achieving something significant. For example, on birthdays, graduations, retirements, weddings, wedding anniversaries, and career promotions, in business you can celebrate mergers, acquisitions, stock growth, or a grand opening.

Social reinforcement: the ultimate culture ingredient is social reinforcement. Social reinforcement is when the culture contains naturally occurring social reinforcements that we encounter every day like smiles, acceptance, praise, acclaim, inclusiveness, and attention from other people. The safety leaders have to ensure that these reinforcements are based on doing the right things, not the wrong things. Remember, you can also reinforce bad or undesired behavior. A consequence of social reinforcement is self-reinforcement.

Self-reinforcement: is when an individual approves of their own behavior because it meets a defined standard or a clear expectation. An individual can judge for themselves if their actions are achieving success. According to Maslow's hierarchy of needs, self-reinforcement would be close to self-actualization.

Feedback: any and all information received during or after a behavior has occurred. Feedback is critically important for performance. Feedback regarding human performance requires context, and here are some guidelines for feedback context.

Guidelines for feedback

- Feedback should never be weaponized to attack someone. When someone is open to feedback, they are vulnerable. Feedback can be damaging if you use it as a weapon to attack people about what they have done wrong or attack their character as a person—it reveals your true intent, which is that you don't like them, and you want to hurt them. Attacks open the door to bullying, prejudice, and even racism (Rojas Tejada et. al., 2011).
- Feedback is about timing—the right words at the right time and from the right person make all the difference in the feedback being effective. On a side note, why do so many people feel they are qualified to give someone feedback?
- Feedback should be provided in humility with a purpose to improve their performance. Unconsciously, people come across condescendingly, or arrogantly when giving feedback. Being unconscious of your own performance, providing feedback, is a serious lack of awareness that disqualifies you from providing feedback. The whole point of feedback is to improve performance—if your feedback does not measurably improve performance, you need to stop providing it! You're not helping.
- Feedback is something people should desire. This is why you should ask first. Coaching is relational, and therefore feedback is based on trust, empathy, and caring, and if an individual doesn't want your feedback, you're doing it wrong!

Now that the context has been established for the mindset of the coach, we can now establish the guidelines for feedback. The following five steps are how to properly provide performance feedback.

Guidelines for performance feedback

1. Ask permission to provide feedback to make sure the person is in a good place to receive your feedback. We are not always ready for feedback, and if not, the feedback will be wasted, and that means the person does not benefit and does not improve.
2. Feedback should be helpful to the person. It should make them better in how they perform. It's sad the amount of feedback given to people is more about the person giving it, than helping the person who is receiving it. Feedback is not about your knowledge, ego, experience, or other boring dribbles that you profusely espouse with your expanded verbal truckage.
3. Feedback should be expressed in a supportive tone of voice to the person so they can receive the feedback. Your intent should be to help the person improve, but that means your tone has to be receivable to that person. A supportive tone of voice does not always mean soft, it can

also be a harsh tone of voice, but the key is the depth of the relationship. Sometimes feedback can sound intense, loud, and critical, but the people involved have a deeply trusting relationship and know that it is passionate feedback. If there is no relationship, then you're just a total jerk.

4. Your feedback should increase the trust this person has in you. Your conversation should increase their belief that you want to help them improve their performance.
5. Feedback should cause positive emotions, not negative emotions in the person receiving the feedback.

Coaching: is a *relational process* that aims on improving the performance of an individual by focusing on the present performance in regard to the level of performance versus a defined goal. Coaching seeks to expand an individual's capacity to function in a way that engages their full potential in order to maximize one's performance. Coaching goes beyond teaching by helping the individual to learn specifically in a particular context. *Coaching is not telling people what to do.* Using the term coaching to cover as a corrective action is unethical because you are deceiving a person to think you want to help improve their capacity to function, help them strive for their full potential, and maximize their performance, but in actuality you want them to conform. In reality, you think you're coaching, but in actuality you are striving for compliance—that's not coaching! Coaching, a relational process, is about reinforcing desired behaviors to repeat towards habit strength, and corrective action is about using a punitive to get a behavior to stop. Don't disguise coaching as corrective action, when you actually just want to punish a person by putting a mark on their record. Otherwise, why do you record the corrective action? Why does it go in a file? Is it so you can build a case that has legal support?

The help of an industrial-organizational psychologist is a critical need in any BBS program because people often confuse behavior with other things like values, goals, principles, wishes, desired outcomes, and other unsubstantiated ideas, but if it cannot be observed (seen or heard), it is not a behavior.

Taking it a step further, safety behaviors must be defined in terms of desired behaviors. One of the biggest mistakes I see repeated over and over in safety programs is that behaviors are not defined, and if they are, they are framed in the negative. Usually in terms of what is not desired. Safety programs create "Do Not Do" lists. Do not do this and do not do that—which establishes a lot of negativities in the program and the safety culture. Negative behaviors are not reinforceable.

It is impossible to reinforce a behavior framed in the negative. As individuals, we learn based on the frequency of associations we make with something, and associations become habits. One of the conditions of the safety culture that business and safety leaders can establish is defining behaviors in terms of desired behavior. Your safety culture should be established on positive conditions.

Chapter 6

Practicing performance coaching

PERFORMANCE COACHING VS. DEVELOPMENT COACHING

Coaching has become popular. Most people have heard of it, and many have experienced some kind of coaching, and so many people refer to themselves as a coach. According to coaching-online.org, there are 31 types of coaching: Life Coaching, Business Coaching, Executive Coaching, Leadership Coaching, Career Coaching, Organizational Coaching, Relationship Coaching, Intimacy Coaching, Personal Development Coaching, Confidence Coaching, Strategy Coaching, Wellness Coaching, Group Coaching, Team Coaching, Success Coaching, Happiness Coaching, Empowerment Coaching, Social Coaching, Mental Health Coaching, Transformation Coaching, Somatic Coaching, Intuitive Coaching, Behavioral Coaching, Personality Coaching, Inspirational Coaching, Life-Transition Coaching, Career-Transition Coaching, Self-Love Coaching, Skills Coaching, Performance Coaching, and High-Performance Coaching. There's also life and spiritual coaching, and a variety of other made-up coaching. Coaching can be divided into two buckets: development and performance. The first 29 types of coaching fit into the development coaching bucket, and the last two types fit into the performance coaching bucket.

Development coaching is a journey or an elongated process that fosters growth, progress, positive change, and/or the addition of physical, economic, environmental, social, and demographic components. It involves a step-by-step process usually with goals and milestones. The purpose of development coaching is to bring about change that allows someone to achieve more of their human potential by engaging in more of their functional capacity to do things effectively.

Performance coaching is the practice of reinforcing defined and desired behaviors to achieve an expectation through the establishment of high-performing habits and drive excellence, and it is focused on the immediate moment and the immediate behavior. The ABC model of behavior (Chapter 5) is implemented with a clearly defined behavioral expectation.

DOI: 10.1201/9781003340799-8

THE SEVEN STEPS OF PERFORMANCE SAFETY COACHING© CONVERSATIONS

Step 1. Ask an open-ended question to start a conversation that helps you understand the performer's perspective

An open-ended question usually starts with Who, What, Why, Where, When, How, How often, and/or To what extent and allows a person to explain their thought process or point of view. The open-ended question allows you to ask and then listen with the purpose of learning. For example, "So what happened in this circumstance, can you please walk me through it?" Another example is "In a situation like this, what is your safety focus?"

There are several types of questions that a leader can use to elicit a more accurate response when engaging in a conversation with an individual. It is important that safety leaders learn what types of questions there are, what kind of response they produce, and when to ask the right type of question to accurately assess information and insightfully use that information to influence people accordingly. The types of questions are as follows: probing, leading, loaded, funnel, recall, process, rhetorical, closed, and open.

Probing questions are useful for gaining clarification about a particular subject and are usually a series of questions that are built off the last question and dig deeper to provide a full picture of understanding. For example, "what do you need to finish the project?" and "can I email you that information later today?"

Leading questions lead the respondent towards a desired logical outcome. For example, "do you have any problems with the next steps?" or "do you see how these next steps are helping you?" The first question prompts a negative response, while the second prompts a positive response. Leading questions can also coerce a respondent into agreeing with the speaker. For example, "I really think this project is going well, don't you think so?" The person is coerced because they are put on the spot.

A loaded question is a closed-ended question with a twist that contains an assumption about the person being asked the question. The loaded question is also known as the trick question because they are worded in a way that can trick the person into admitting something that is true or may be false. For example, "have you stopped stealing my ideas?" The assumption is that the person stole your ideas, and they have done this before. It is a manipulation technique.

A funnel question starts broadly and then narrows to a specific point. For example, police may ask a witness, "where were you the night of the 24th?" "what were you doing there?" "how long did you stay?" "did you see the blond woman with the purse?" This also works vice versa. For example, when making introductions you may ask specific closed questions, "what's your name?" "What do you do for work?" and as you get more comfortable you broaden your questions "what got you into engineering?" Funnel questions can also be used to diffuse a tense situation. For example, a hostage negotiator will use funnel questions to keep the person talking and learn more about the situation. When talking to a suicidal individual, funnel questions help the person process their thinking. Asking someone to explain in more detail helps them to focus on something they can get their head around and helps to change their perspective and understand a possible solution that helps them calm down.

A recall question involves the remembering of facts. For example, "when did you last log in?" "What time did you leave?"

A process question asks a person to add their opinion to a previous answer. This type of question can test an individual's depth of knowledge about a topic. "What would you add as a next step?"

A rhetorical question is interesting in that it does not require an answer and does not require any real mental processing because it is easy to go along with. It is a statement phrased as a question to engage the listener, and it draws them into agreement with you. For example, "It is so nice working with talented people, isn't it?" "Having a great salary is what you want, right?" The rhetorical question is a tool to get someone to think and agree with you.

Closed-ended questions invite a one-word response, most often a yes or no, and are used to gather facts objectively. For example, "do you come here often?," or "did you finish your coffee?"

Asking meaningful questions is a crucial part of the safety culture. Meaningful questions provide your safety people permission to speak freely and be curious and creative. They learn to think and ask questions in ways that help them become better critical thinkers. They also get to strive for deeper knowledge and more meaningful answers.

Step 2. Practice active listening to understand the performer's experience, perspective, and effective potential consequences

Active listening is a technique that keeps you, the listener, engaged and attentive while someone else speaks. Paraphrasing and reflecting back to them what the person is saying, seeking clarity, and understanding the information being shared. It is a positive experience, and the process is attentive, focused, and concentrated on comprehending the information presented and how it fits with other information.

Step 3. Express empathy when an opportunity is presented, to connect with the performer

Properly expressed empathy connects us with another individual by considering another person's situation or circumstance. However, 99% of people express apology to others and not empathy. Saying, "I'm sorry for your loss" is an apology. An apology is defined as a regretful acknowledgment of an offense or failure. Empathy is the capacity to express what another human being is experiencing from within the other person's frame of reference. Saying "I'm sorry" does not accomplish the full requirement of empathy (i.e., the capacity to place oneself in another's position).

Empathy requires a capacity to understand or feel the other person's experience, and how that is expressed is crucial to successful empathy. The key to empathy is to express to the individual what you see in their face or hear in their voice. It is an honest expression and an authentic expression based on the experience of that moment. "I can hear the frustration in your voice, and I would be frustrated too. Let me help you get this situation resolved." "I see you are upset; do you want to talk about it?" You must honestly and authentically capture the moment to be empathetic. Most people think empathy is an apology. I'm not sure why, but most people say they are "sorry" when something bad happens to someone. An apology misses the impact of empathy, and technically it is an admission of guilt, which is very odd. The Feeling Type personality is the most natural at empathy; however, it is a learned behavior, and therefore anyone can become skilled at empathy. Sympathy is having feelings of pity or sorrow for those in need and is not as intimate or artistic as empathy.

Step 4. Identify and clarify the root issue of the behavior to improve with a clearly defined behavior and examine for any "hidden voice" if the opportunity was presented

There are many process-focused methodologies, approaches, and techniques for conducting a root cause analysis, including Five Whys, Events and Causal factor analysis, change analysis, Barrier analysis, Risk Tree analysis, and Kepner-Tregoe Problem Solving and Decision Making. However, you need to know how a behavior is defined, what prompts a behavior, and what motivates (i.e., consequences) a behavior. Understanding the ABC model of behavior is a key underpinning at this stage of the coaching conversation. Human performance is often limited due to safety and business leaders not understanding the fundamentals of human behaviors and the ability to influence improved performance. Safety leaders must be well versed in identifying which of the six consequences is being engaged. Identifying the consequence accurately is the core skill of the Safety Performance Coach©. In addition to consequences safety leaders need to consider environmental factors, social norms, personality, and habits as antecedents that prompt behaviors to occur.

Step 5. Gain agreement with the performer on pinpointed behavior to establish accountability

Gaining agreement is getting the performer to commit to an obligation regarding a specific and a clearly defined safe behavior. It is about influencing the performer to express a willingness (i.e., discretionary effort) to accept responsibility to act or behave in a way that is clearly defined as safe. The experience of this interaction is premised on positivity influencing and agreeing to conduct or perform towards something good—a clearly defined safe behavior, and something achievable.

Step 6. Ask the performer how they will improve their performance going forward and establish the next steps on what the performer will do to improve

Asking the performer, based on the agreed behavior, how they will improve engages them to think about their own safe behavior. This is what Behavior-Based Safety (BBS) is all about, reinforcing people to think safely. Asking the performer how they will improve gets them to think critically about their future performance and how they need to conduct themselves. This is also why it is called performance coaching because the focus is on performing safely, right now. Performance is defined as the execution of an action, something accomplished.

Step 7. Ask the performer to give you feedback about the coaching conversation you just had with them

Asking for feedback is a strong demonstration of openness. For example, "How was this conversation for you?" "What was helpful to you about this conversation?" Being open to the performer's feedback reinforces them to be open to your feedback. It is crucially important that feedback be experienced as a positive norm. A Performance Safety Coaching© conversation should be about the desired behavior, encouraging the performer to adjust and do better, and that you as the coach are supportive of that effort to improve.

All Performance Safety Coaching© conversations should be recorded as data, and each follow-up conversation should have its data recorded, until the behavior is tracked to habit strength. This is critical to any BBS program.

DEMONSTRATING FLUENCY OF THE SEVEN STEPS OF PERFORMANCE SAFETY COACHING©

Fluency is the ability to utilize the seven steps of the Performance Coaching conversation in an easy and smooth fashion. It is the confidence to move through the steps in a way that is supportive of the performer and yet in a way that improves their performance. The Performance Coach needs clearly defined expected behaviors to perform to the expected level of coaching:

- Provide feedback in a way that helps the performer improve their behavior. The Performance Coach must express an objective tone of voice that is factual and actually helps the performer improve. You cannot just offer anecdotal (i.e., based on unscientific knowledge) advice that is not measured or measurable.

- Speak with a positive tone of voice that helps the performer receive the feedback. BBS requires a supportive premise. The coach's motivation is to help the performer improve. Your style can be direct, philosophical, intellectual, fun, etc., but it must be supportive and focused on a clearly defined desired behavior. The bottom line is that you need this person on your team, and you need everyone on your team to perform to the required level that achieves the goals.
- Focus on a positive behavior that you want to create an opportunity to provide positive reinforcement to the performer frequently. Focusing on the negative does not allow for reinforcement. Undesired behavior and reinforcement are opposites. Reinforcement gets a behavior to repeat, you get more of that behavior. It repeats itself due to the reinforcement. Focusing on negative behaviors shifts from coaching to "telling." Telling people what to do requires no thinking on their part. Telling people what to do comes across as authoritative and can be condescending and reinforces a negative culture. If someone does something wrong, start with step one of the performance coaching.
- Notices and reinforces incremental improvement in the performer's performance. When an individual is learning a new behavior, they need lots of reinforcement. A good practice is to catch them in the act, let them know they are doing it right, and keep at it. The effort here is to get the desired or required behavior to habit strength.
- The coach does what he/she says he/she is going to do resulting in a trusting relationship. Trust is always earned by doing what we say we will do. Trust is earned by doing the little things. For example, returning a call when you said you would. Emailing the information that you said you would. When you make small commitments, and you keep them you earn people's trust. Doing the small things is part of a much larger thing, and that is your character. Your character is made up of the mental and moral qualities that are distinctive to you as an individual.

PRACTICING FREQUENCY OF REINFORCEMENT TO SHAPE BEHAVIOR TO HABIT STRENGTH

A habit is a settled tendency of thinking that allows for a learned mode of behavior to be automatic (i.e., seemingly natural) in its practice. Shaping a habit takes lots of perfected repetitions and frequent positive reinforcement. The ability to reinforce a behavior until that behavior is at habit strength is essential to sustaining continuous improvement. Habit formation is a process by which a behavior is perfectly repeated and reinforced properly (i.e., punishment to penalty, penalty to negative reinforcement, negative reinforcement to positive reinforcement, positive reinforcement to a sustained habit of behavior) until it has become automatic or habitual (i.e., habit strength).

You have undoubtedly heard that it takes 21 days to form a habit, but this is fake science. This concept of 21 days to form a habit was introduced by plastic surgeon Maxwell Maltz in 1950. He based his idea on the fact that it took a patient 21 days to get used to the change of their face surgery. You may ask, "how does that relate to a habit?" It does not, and now the world thinks it takes 21 days to shape a habit. Research shows that it can take anywhere from 18 to 254 days for a person to form a new habit, which is an average of 66 days.

It is the reinforcement of that behavior that shapes the behavior to habit strength. When you change a behavior, you have a different neural connection, and therefore you need to incorporate reinforcement practices to get that behavioral pattern to habit strength.

People often confuse reinforcement and reward, but the two are technically different and need to be practiced accurately to get the desired result. Reinforcement will "carve" the new pathway in your brain and establish the pattern to habit strength, while the old path will disappear.

Therefore, if you teach people how to think safely and reinforce people when they think safely, the frequency is the key that makes it become a habit, and then it becomes the norm (if the reinforcement continues until habit strength is achieved). It becomes a pattern in their brain, and the individual behaves accordingly. The behavior must be clearly defined, antecedents must be used appropriately to remind or prompt the clearly defined behavior, and the consequence has to be a positive or negative reinforcement to maintain a repetition of that behavior.

The ABC model of behavior helps us understand how to shape a behavior to habit strength. The first step is to clearly define behavior. A behavior is defined as something you can see or hear. People confuse behavior with lots of other terminologies, but this is the reason a definition is needed. In science, every term needs a strict definition, and you must use that word strictly within that definition. Now that we understand the definition of behavior, it is important that you define a desired safety behavior as positive, something expected. Once the behavior is defined, you need to create antecedents in the safety culture that help prompt or remind people to perform the expected behaviors. Remember, an antecedent comes before the behavior and only prompts or reminds someone to do the behavior. Antecedents do not cause behavior. Consequences cause behaviors (to repeat or stop). Putting up a poster is an antecedent, and it is not enough. A poster or sign will not cause a behavior to occur, it only reminds people, and it should focus on a behavior, not a goal, value, vision, or something else. People will ignore a poster or a sign for many reasons. You can put a sign right in front of someone and they can choose to ignore it. Antecedents do not cause a behavior to occur.

Consequences cause behaviors to repeat if it is a reinforcement type of consequence. A punitive type of consequence causes a behavior to stop. It is critical that safety leaders understand how consequences work because this is what causes a behavior to happen. It is also important that safety leaders create conditions in a culture that influence people through reinforcement (i.e., positive, negative, or recovery), and not a culture of punitives (i.e., punishment, penalty, or extinction). Most safety leaders fill the rooms and hallways with colorful posters, signs, commitments where employees sign their names, provide training, workshops, speakers, establish safety meetings, protocols, and procedures. None of these things are wrong, but they all fall under the category of the antecedent. They are prompts or reminders to do a thing. Antecedents do not cause a behavior to occur. The questions I like to ask myself when analyzing a safety area are what are they reminding their workers to do? What is the required behavior(s)? I see the phrase "be safe" a lot. What does "be safe" mean? It is not a clear message because it is not a defined behavior. What is the desired behavior(s)? Safety leaders get this wrong all the time. The intention of safety leadership is in the right place, but they get it wrong, badly.

Chapter 7

Safety campaigns should be behavior campaigns

Traditional safety campaigns are confusing. The confusion starts with complicated themes and the wrong focus. Traditional safety campaigns use banners, slogans, posters, training, and safety committee meetings to reinforce one-year or two-year-long themes. This is too long. If you can't read the slogan and understand it instantly, then the campaign is a failure. And yet for the next year or two years, this failure of a campaign will be pushed ineffectively by the safety leadership team.

Here are some famous campaign slogans:

> "Safety is no accident."

This slogan is a play on words, but it is confusing. What is the behavior that people should do? How should someone think and act safely? This comes across as cute, and not a demonstration of effective safety leadership. There needs to be a metric tied to campaign slogans so that they are meaningful and aligned to the vision, purpose, values, and measurements of the safety program.

> "Stop! Think! Then act!"

This slogan is behavioral focused, it is specific, but it is telling you what to do and is a practice of Results-Based Safety (RBS), not Behavior-Based Safety (BBS). The intent is for people when they see a dangerous situation to not react instantly and instinctively (negative focus). What the safety leadership wants is for people to approach any situation with a reasoned plan, based on training and situational awareness (desired behaviors). Unfortunately, they do not say that. This is an example of a safety high-impact safety slogan, but there are no metrics to prove high impact.

> "Leave sooner, drive slower, live longer."

This slogan may be true, to a degree, but it promotes a false (not to mention super negative) outcome. The intent is well-meaning, but the science is

DOI: 10.1201/9781003340799-9

missing. Encouraging employees to prepare for any work-related departures as early as possible is the desired safe behavior. This is a fear-based premise because safety leadership is thinking that they don't want employees to find themselves running late and rushing, which might lead them to exceed the speed limit or disregard any road rules (not much faith in your people). The solution is to tell them what to do—problem solved, but is it? This is RBS, not BBS.

"Your good health is your greatest wealth."

This slogan has a wonderful meaningless message. Is this statement even true? This is not a behavior. It is assumed that everyone wants to maintain good health and safety in the workplace and at home. Safety and business leaders will talk about work-life balance and maybe provide incentives, such as flexible schedules, paid time off, or sponsor morale-building activities. However, what are the behaviors? Once again, this slogan misses the point completely.

"Be aware, take care."

How does one "be aware"? This slogan is not clear on what one should do. Because there are risks that abound in the workplace, you need to stay aware of your surroundings. The idea here is to become conscious of each risk and help your team members remain mindful. However, this is an antecedent and not a behavior. It is a good reminder to help your fellow employees to stay aware and alert, but an antecedent has little impact on behavior. Safety leadership has to define the behavior that people need to practice.

"A spill, a slip, a hospital trip."

What is the intent of this slogan? Nobody wants employees to experience an injury that leads to a hospital trip. The point of this slogan is to encourage everyone to use caution and care in the workplace, and if someone accidentally spilled coffee in the breakroom that could lead to a slip, awareness is crucial. Does this make sense to you because it doesn't make sense to me, especially when sounding cute about going to the hospital for an injury? Awareness will not keep people safe, cleaning up the spill will keep people safe.

"Never give safety a day off."

The whole point of this slogan is the wrong focus. Safety doesn't take days off—a confusing message because the focus should be on people. What the safety leadership team means is that any time a worker is on the organization's premises, hazards are a fact of life. The thinking here is to offer reminders that help workers stay mindful and careful. Again, this is an antecedent—a prompt or a reminder to do a behavior, but what is the behavior?

> "Think safety—it couldn't hurt."

Why the sarcasm with this slogan? It comes across as if the safety leadership team is frustrated. Their frustration is a sign that their Results-Based Safety is not working. This slogan intends to reinforce the idea that the safety leadership team's primary objective, and yours, is to think about known risks in the workplace and to avoid a painful accident or injury. Unfortunately, this statement tells you what to think. Telling people what to do does not work. This slogan is meant as a reminder, an antecedent, it is not meant as a behavior, but thinking is a behavior. They actually get this right, but they don't know it because they don't understand the science of behavior. Safety leaders continue to confuse antecedents with behaviors.

> "Don't be a fool—use the proper tool."

This slogan is derogatory and unprofessional. It seems that safety campaigns continually focus on being cute. We all understand that we need to use workplace tools according to their intended purpose to ensure safety. Even if you have lost one tool, they do not want you to improvise with something that can serve as a substitute. For example, don't use a monkey wrench to hammer a nail if you can't find a hammer. If a specific tool you need has gone missing, work with your supervisor to find the proper tool or equipment before proceeding. There are a lot of behaviors that could be focused on here, but they are passed over. Safety leaders need to focus on behavior intentionally, but they continue to miss the mark.

> "Accident prevention—your no. 1 intention."

This is a generalized slogan that clearly focuses on a result, this is not BBS. It is catchy, but it is meaningless. Accident prevention is a goal, and therefore Results-Based Safety, not Behavior-Based Safety. Safety leaders want to demonstrate that they care about their employees' safety, health, and happiness as key ingredients for their well-being and ongoing productivity, profits, and brand reputation. Focusing on preventing accidents is a good thing, and it is beneficial to everyone and should be your primary goal. But it is a goal, not a behavior.

Safety leaders continue to demonstrate that they do not understand the science of human behavior in their safety campaigns. A safety campaign focused on results is Results-Based Safety in action. When safety leaders do not achieve results, they get upset. The business leader puts the pressure on the safety leader to get results, and this is where things go wrong.

Safety campaigns need to be designed as a series of strategic advertising messages that share a single idea and theme which make up an integrated marketing communication of safe behavior. The safety campaign should be a comprehensive course of action to promote safe thinking through different types of media, such as television, radio, print, and online platforms.

Chapter 8

Evidence of a healthy Behavior-Based Safety program

A healthy Behavior-Based Safety (BBS) program provides several benefits to the employees, management, and leadership that are critical indicators that you are developing your people properly. These include locus of control, self-esteem, self-efficacy, and self-actualization and develop the capacity for fully functioning individuals and teams. This is the ultimate outcome of a BBS program—to develop fully functioning leaders, individuals and teams.

LOCUS OF CONTROL

Understanding locus of control can have significant impact in differentiating between effective and ineffective leadership, management, and employee performance. Locus of control is a psychological concept that refers to how strongly people believe they have control over the situations and experiences that affect their lives. Locus of control is the degree to which people believe that they, as opposed to external forces, have control over the outcome of events in their lives. The concept of locus of control was developed by Rotter (1954) and plays a significant role in safety. Every individual needs to believe they have control over the situation they are in through planning, preparation, training, and reinforcement to such a degree that they fully believe in their abilities to conduct themselves safely.

The importance of locus of control is based on meta-analysis research. Based on 135 research studies "internal locus of control was associated with higher levels of job satisfaction and job performance" (Colquitt, LePine, & Wesson, 2015, p. 287). A second meta-analysis of 222 research studies showed that "people with an internal locus of control enjoyed better health, including higher self-reported mental well-being, fewer self-reported physical symptoms" (Colquitt et al., 2015, p. 287). Every individual needs to take accountability for their own safety and the safety of others. Safety leadership needs to create a culture that helps develop every individual's capacity to be safe.

DOI: 10.1201/9781003340799-10

There are two types of loci of control: internal (inside) and external (outside). Internal locus of control is the belief that you are "in charge of the events that occur in [your] life" (Northouse, 2013, p. 141), while external locus of control is the belief that "chance, fate, or outside forces determine life events" (p. 141). Individuals with an internal locus of control believe their behaviors are guided by their personal decisions and efforts and they have control over those things they can change. Having an internal locus of control is linked to self-efficacy, the belief you have about being able to do something successfully (Donatelle, 2011). People with an external locus of control see their behaviors and lives as being controlled by luck or fate. These individuals view themselves (i.e., their lives and circumstances) as victims of life and bad luck.

> People differ in whether they feel they control the consequences of their actions or are controlled by external factors. External control personality types believe that luck, fate, or powerful external forces control their destiny. Internal control personality types believe they control what happens to them.
>
> (Champoux, 2011, p. 113)

When investigating a negative safety incident, if you really listen, you'll often hear an individual describe their experience as if they felt they had very little control over or were not in control of the situation. However, when coaching someone during a positive safety event, you hear them say that they felt in control. They knew what they were doing and were confident in their planning to be safe. When the locus of control shifts from the external to the internal frame, employees find more energy, motivation, and greater confidence to continue to improve their behavior (Moore & Tschannen-Moran, 2010, p. 75).

Safety leaders can realize the benefit of having an internal locus of control and make it applicable to all individuals at all levels within their organization:

1. An internal locus of control is one of the key traits of an effective leader (Yukl, 2006).

> A leader with an internal locus of control is likely to be favored by group members. One reason is that an 'internal' person is perceived as more powerful than an 'external' person because he or she takes responsibility for events. The leader with an internal locus of control would emphasize that he or she can change unfavorable conditions.
>
> (Dubrin, 2010, p. 47)

2. An internal locus of control separates good from bad managers (Yukl, 2006).

Effective managers . . . demonstrated a strong belief in self-efficacy and internal locus of control, as evidenced by behavior such as initiating action (rather than waiting for things to happen), taking steps to circumvent obstacles, seeking information from a variety of sources, and accepting responsibility for success or failure.

(Yukl, 2006, pp. 185–186)

3. Employees' locus of control affects leadership behavior in decision-making (Hughes, Ginnett, & Curphy, 2012).

Internal-locus-of-control followers, who believed outcomes were a result of their own decisions, were much more satisfied with leaders who exhibited participative behaviors than they were with leaders who were directive. Conversely, external-locus-of-control followers were more satisfied with directive leader behaviors than they were with participative leader behaviors. Followers' perceptions of their own skills and abilities to perform particular tasks can also affect the impact of certain leader behaviors. Followers who believe they are perfectly capable of performing a task are not as apt to be motivated by, or as willing to accept, a directive leader as they would a leader who exhibits participative behaviors.

(Hughes, Ginnett, & Curphy, 2012, pp. 544–545)

There is also evidence that internals are better able to handle complex information and problem solving, and that they are more achievement-oriented than externals (locus of control). In addition, people with a high internal locus of control are more likely than externals to try to influence others, and thus more likely to assume or seek leadership opportunities. People with a high external locus of control typically prefer to have structured, directed work situations. They are better able than internals to handle work that requires compliance and conformity, but they are generally not as effective in situations that require initiative, creativity, and independent action.

(Daft, 2008, p. 103)

Path–goal theory suggests that for employees with an internal locus of control participative leadership is most satisfying because it allows them to feel in charge of their work and to be an integral part of decision making. For employees with an external locus of control, path–goal theory suggests that directive leadership is best because it parallels employees' feelings that outside forces control their circumstances.

(Northouse, 2013, p. 141)

SELF-ESTEEM

Self-esteem impacts one's decision-making process, relationships, emotional health, and overall well-being. It also influences motivation, as people with a healthy, positive view of themselves understand their potential and potentially feel inspired to take on new challenges.

Individuals who are more confident in their ability to make decisions are able to form secure and honest relationships and are less likely to tolerate unhealthy ones. High self-esteem helps people to be realistic in their expectations and less likely to be overcritical of themselves and others, and they are more resilient and better able to weather stress and setbacks (Williams, 1999). High self-esteem is important to a healthy safety culture and supports the empowerment of people who build self-confidence and increase assertiveness in the context of feeling safe. A key element of self-esteem is empowering those who have been marginalized, excluded, or denied a voice. Safety needs to include diversity and inclusion of all the employees, and the development of self-esteem supports this objective.

SELF-EFFICACY

Self-efficacy refers to an individual's belief in his or her capacity to execute behaviors necessary to produce specific performance attainments (Bandura, 1977a, 1986, 1997). Self-efficacy reflects confidence in the ability to exert control over one's own motivation, behavior, and social environment. These cognitive self-evaluations influence all manner of human experience, including the expectations set by management, the goals for which people strive, the amount of energy expended towards goal achievement, and the likelihood of attaining particular levels of behavioral performance within the safety environment.

An individual's belief in their capacity to execute behaviors is critically important and necessary to produce specific performance expectations. Safety leaders need to think about how they define requirements and how those requirements are linked to the safety goals of the organization. Safety leaders need to consider the impact of their expectations and how they are integrated to achievement. Self-efficacy affects every area of human endeavor, and safety leaders need to support the development of every individual's self-efficacy.

SELF-ACTUALIZATION

Self-actualization is based on Maslow's hierarchy of needs. It is the highest level of psychological development, where personal potential is fully realized. Self-actualization is the realization or fulfillment of one's talents and potential. Safety leaders have to shape conditions in their safety culture to help people strive for their self-actualization. The beauty of a BBS program is that the capacity for individual potential to achieve self-actualization is possible, and even more so in an organic culture. Any safety leader can meet the basic needs of their people, such as safety, belonging, and self-esteem, and create cultural conditions in order to satisfy (to a reasonably healthy degree) people's needs so that they can be able to fully realize one's unique creative and humanitarian potential.

FULLY FUNCTIONING INDIVIDUALS AND TEAMS

Fully functioning individuals do not live in fear. A fully functioning person makes the best employee. They have a healthy personality because they experience the freedom of choice and freedom of action, and they are creative and exhibit the qualities you are shaping in the safety culture. Rogers (1959) stated that for a person to "grow," they need an environment that provides them with openness and self-disclosure, acceptance (being seen with unconditional positive regard), and empathy (being listened to and understood). Empathy is a skill that is seriously lacking in safety leaders, and yet it is one of the most significant qualities of a safety leader.

The safety leader has to shape the safety culture by creating conditions that foster relationships and healthy personalities, growth and development, and the highest performance possible from their people every single day. Safety leaders should believe that every person could achieve their goals and behave safely in every situation. Self-actualization is the realization of an individual's potential, but that requires a perceptual schematic.

PERCEPTUAL SCHEMATIC

A safety culture needs a perceptual schematic to help members of the culture to understand the vision, purpose, values, and performance goals. The safety campaign helps to reinforce safety perceptions. A perceptual schematic is a mental model that provides a frame for interpreting information entering the mind through the senses or for activating an expectation of how particular safety behaviors or situations may look like.

The perception process is how we mentally arrange stimuli from our environment into meaningful and comprehensible patterns. Safety leaders have to be more intentional in how they develop their safety culture. Safety leaders have to link purpose with planning and planning with the reinforcement of required and desired safe behaviors. The safety leader needs to connect people and their perceptions constantly and consistently to the safety culture. The safety leader's responsibility is to create and lead a functional safety program.

Part 3

Structuring the culture for functional safety

Chapter 9

Designing the safety culture on purpose

Safety culture design is a step-by-step methodology, which envisions aspects of workflow, procedures, structures of responsibility, and communication systems and aligns them to fit current business realities and achieve goals, but also with an eye to implement the new changes to stay current with demands of customers, society, or the local and federal government as well as the process that focus on improving both the technical and people side of the business.

For most companies, the design process leads to effective performance through consistently improved results (profitability, customer service, internal operations), and employees who are empowered and committed to the business. The hallmark of the design process is a comprehensive and holistic approach to organizational development and improvement that touches all aspects of organizational life, so you can achieve excellent service, increased effectiveness, reduced operating costs, improved efficiency and cycle time, a culture of committed and engaged employees, and a clear strategy for managing and growing your safety influence.

Safety culture design is defined as a look and functioning of a safety culture, by making a detailed drawing of it. By design, we are talking about the integration of people with core safety processes, technologies, and systems. A well-designed safety culture is designed to ensure that the formation of the environment matches its purpose and strategy, meets the challenges posed by business realities, and significantly increases the likelihood that the collective efforts of people will be successful.

As companies grow and the challenges in the external environment become more complex, business processes block improvement and organizational structures and systems that once worked become barriers to efficiency and thus impact employee morale and financial profitability. Safety programs that do not periodically renew themselves suffer from such symptoms as an inefficient workflow with breakdowns in valued processes, non-value-added steps, redundancies in effort ("we don't have time to do things right, but do have time to do them over"), fragmented work that has no regard for the cohesion of the whole workplan, lack of knowledge,

DOI: 10.1201/9781003340799-12

inaccurate communications, silo mentality, turf battles, and lack of ownership ("It's not my job") and cover up or blame rather than identifying and solving problems. Any delays in decision-making, lack of information, misinformation, or lack of authority to solve problems when and where they occur, is the responsibility of management—rather than the front line. Management is responsible for solving problems when things go wrong, but when management practices "telling people what to do," it takes a long time to get something done, systems are ill-defined or reinforce wrong behaviors, and there exists mistrust between workers and management. It can become an ugly spectacle when the safety culture is not controlled purposefully by the safety leader.

One of the biggest indicators of an unhealthy safety culture is the lack of transparency through the practice of omission. Transparency is crucial to culture and is demonstrated when it is easy for others to see what issues are out there, what decisions are being made, and what actions are performed. Omission undermines trust in the safety leader, the leadership team, and is a threat to the culture. Omission occurs when information or actions are left out or excluded, or when someone fails to act according to any moral or legal obligation to do so. Leaders must understand that trust is a result, a result earned from acting accordingly and being transparent. Omission and lack of transparency are underpinnings that de-motivate people within the culture. Motivation includes psychological constructs like locus of control, self-esteem, self-efficacy, and discretionary effort. It is critically important that the safety leader knows how to design the safety culture on purpose, and with care.

ESTABLISHING SAFETY CULTURE LEADERSHIP

Business and safety leaders must understand the current conditions that make up their culture and plan to create better conditions that align with all performance criteria. A condition is defined as (a) the state of something with regard to its appearance, quality, or working order, (b) the circumstances affecting the way in which people live or work, especially with regard to their safety or well-being, (c) have a significant influence on or determine (the manner or outcome of something), or (d) bring (something) into the desired state for use.

One type of condition are phenomena, which are events that have been observed and considered factual (*actuality*). However, the cause or explanation is elusive and therefore difficult to understand (*reality*). In psychology, phenomena consist of frequently observed (seen or heard) behaviors in a systematic empirical structure. This is important because phenomena are underpinnings to the environment as we experience it in our individual *reality*, and not necessarily as it *actually* exists.

The word *actual* is used intentionally to differentiate from *reality* as every individual's *reality* is their own. Every individual has a skewed reality, and that reality does not mean it is *actually* as it is, because individuals have skewed perceptions of what is actually happening. *Actuality* is defined as existing in fact, contrasted with *reality* which is defined as what was intended, expected, or believed. The safety leader must have the competence to discern the difference between reality and actuality to properly influence people towards the desired state they want to achieve in their safety program.

This duality of *reality* and *actuality* plays out within cultural phenomena. Safety leaders must think through and realize the extent of the complexity of conditions (e.g., policy, procedure, SOPs, communications, leading indicators, lagging indicators, timing, scaling, punitives, and reinforcements) that shape behavior and how they intentionally need to shape behavior towards the required outcome.

This would also include the time it takes to shape behavior, how to shape specific behavior(s), human potential, capacity, and capability, the negative impact of incidents and injuries, the negative impact of *telling* people what to do, and the complexity in coming to an understanding of how phenomena impact the entire culture.

The organizational culture influences the safety culture in the way people interact, the context within which knowledge is created, any resistance that arises towards certain changes, and ultimately the way communication of information is shared (or the way it is not shared). Organizational culture represents the collective vision, purpose, values, beliefs, principles, and personalities of organizational members. It may also be influenced by factors such as history, type of product, marketplace, technology, strategy, type of employees, and management style. Culture includes the organization's norms, systems, symbols, language, assumptions, environment, location, and habits. The safety leader has to be competent in shaping the safety culture, being influential at multiple levels, effective in communication, and driving the reinforcement practices of every individual to practice safe behavior.

Hiring the right safety leader

The first thing to look for in hiring the right safety leader is not their experience, but their competency. It should be the first thing you look for in any significant leadership position, but it is not. Unfortunately, most business leaders look for experience as the highest priority, and based on experience, competency is assumed.

Experience is defined as the direct observation of or participation in events as a basis of knowledge. It is the fact or state of having been affected by or gained knowledge through direct observation or participation in specific things. Experience is something personally encountered, undergone, or lived through. Experience involves the process of directly perceiving events or reality. Experience has its value, but the problem with experience is that it is limited in its capabilities and therefore should not be the major basis for hiring. Competency is a much better thing to seek in a safety leader.

Competency is the modern word for wisdom. Wisdom is an expression of competence and is the ability to demonstrate insightful understanding (sagacity), make good judgments (sapience), and is developed around a particular group of tasks underpinned by one's ability to think and act using knowledge, experience, and the capacity to be self-aware with rare sense and insight. A competency is the ability to do something successfully or efficiently on a larger scale—like create a culture that functions with a high rate of desired safety outcomes. A competency must be developed over time and usually requires complex experiences that include key behaviors in successful planning, measurement, performance, and employee engagement in safety. Competency can be measured, and it would be worth the effort to consider assessing safety competence as part of the interviewing process.

Safety culture is a competency. The safety leader needs to be competent in creating a functional safety culture by designing the safety culture on purpose. This is a significant competency for any high-level safety leader, and yet it is exceedingly rare. Business leaders rarely think in terms of culture, or culture as a competency. The safety culture competency should be part of the interview process, but business leaders do not think of safety as a competency and rarely interview for this. The business leader is responsible for the conditions and the phenomena that sustain a culture, including safety culture, and therefore should interview to examine for this competency. Safety culture is underpinned by competency. The safety culture competency has several components.

The role of the business leader must help and support the creation of an infrastructure that focuses on the safety leader's behaviors and that they are aligned to the organization's core competencies, values, purpose, and vision and then linked to the required outcomes that drive defined success in a way that is harmonious. It is the business leader's responsibility to be aligned with the safety leader. For the safety leader, there are leadership competencies as well as the culture competency. For example, leadership competencies would include clear thinking, problem-solving, use of language (i.e., tone of voice, inflection, communication via verbal, emails, and texts), managing ones-self (i.e., emotional intelligence), relating well to others, and how one participates and contributes to the betterment of the safety culture. The competency of safety culture would include knowing how to clearly define the vision, purpose, and values of the safety culture. Clearly defining the role of the safety position, safe behavior(s), defined expectations, required outcomes, the intent and purpose of policy and procedure, chain of command, chain of communication, reinforcement of what is going right and defining overt and covert behaviors, rewards for going above and beyond, and appropriate recognition for accomplishments of a high order.

The safety leader who demonstrates the competency of safety culture requires having the right blend of experience, knowledge, ability, attitude, skill, and industry-specific insight (KAASISI), which demonstrate a good fit with the business leader, the employees, and the potential capability to accomplish the expectation of the business leader. The safety leader must have a firm understanding of the business strategy so as to align safety with the business. Safety should not be a priority. Safety should be a core value that is part of the organizational culture, which is significantly critical to a successful Behavior-Based Safety (BBS) program.

Business leaders have to be careful to require this expectation for their safety leader to be qualified to lead. Business leaders miss this aspect regularly because (1) business leaders do not understand the power of positioning a safety culture towards the required outcome. (2) Their focus is limited to results only. This is the ultimate root cause for all safety program failures (Results-Based Safety [RBS]). It is a vicious cycle and a sign of weak leadership when a business leader continues to foster the focus on results only. A safety leader should have a different focus—supporting team members through active monitoring of decisions and actions and ensuring alignment with the corporate safety strategy, vision, and values.

Selecting safety leadership competencies

Here are seven competencies that are needed to establish the competency of safety leadership: (1) ability to clearly define the safety competencies of the safety leadership team based on knowledge, ability, attitude, skills, and industry specific insights (KAASISIs) and clearly list the desired and required behaviors for these competencies. (2) Capability to review the current corporate culture and safety culture for gaps and determine what conditions need to be added to achieve a harmonious, reinforcing sustained culture. (3) Create a clear definition of the desired and required behaviors of the safety culture. (4) Establish clear linkages to the corporate culture, thus creating a clean, clear alignment of safety culture with corporate culture. (5) Write job descriptions for the safety leadership positions. (6) Classify a position profile. (7) Develop structured recruiter level interview questions that focus on experiences (to reveal behaviors that you are looking for), write out VP, Director, Manager, Supervisor, and Team Lead interview questions for each level that are scenario based and frame the current situation, then allow the candidate to explain: what steps he or she would take to improve the situation, and the timeline that they expect to accomplish the requirements.

Do not ask the candidate about the timeline. If they do not provide specific answers to your specific questions, then you obviously did not create rigorous enough questions and therefore cannot properly measure the candidate for fit, or the candidate is revealing that they are not a good fit. This process is about measuring for fit and finding the right person to perform in a way that accomplishes the requirements of the position. A list of safety leadership competencies would include:

- Recognizing and reinforcing team members based on their demonstration of effective safety behavior.
- Actively caring for the health, safety, and general well-being of team members.
- Collaborating, or sharing ownership of safety with team members by asking for their active participation in safety decision-making and empowering everyone on the team to take personal responsibility for safety.
- Sharing a vision for safety and facilitating the development of team goals, targets, and plans to achieve it.
- Inspiring the team to achieve the safety vision and safety excellence through motivational and encouraging communications.
- Role modeling safety compliant behaviors that set the benchmark of what is expected from the team.
- Challenging team members to think about safety issues and scenarios in ways that they might not have considered before.
- Positive reinforcement of desired safe behaviors through improved employee productivity, quality, and engagement. Effective safety leadership increases discretionary effort.
- Producing safety leaders by developing their safety leadership abilities in measurable metrics enables them to capitalize on their strengths and develop their areas of opportunity. Learn more about assessing your safety leadership.

Utilizing personality traits and strengths

If you want to expand your ability to influence yourself and others, you need to understand how you are hard wired, your personality traits. Understanding your traits helps you more accurately define your strengths and potential blind spots. Understanding your strengths and blind spots helps you to be more open to feedback, and it helps you to better understand that feedback. You cannot grow your abilities yourself or others without feedback. Think of it this way, if you are an individual who accepts limited feedback, then your influence is limited, and so is your personal growth. *The amount of feedback you can receive is equal to the personal growth you can experience.*

Much of human behavior is based on functioning and attitudes. The concept of psychological functioning is related to the way in which our thinking interacts with our feeling and behavior. It has to do with how our brain (physiological) and mind (psychological) are working together, and how our personality contributes to the development, performance, and maintenance of our functional well-being. Functioning is affected by attitude.

Attitude

An attitude refers to evaluations of ideas, events, objects, or people and involves the shaping of emotions and opinions towards a particular object, person, thing, or event. An attitude becomes a settled way of thinking or feeling about someone or something, typically one that is reflected in a person's affect—their mental tone. Attitudes are often the result of perception and can have a powerful influence over behavior. The ABC tripartite model of attitudes, not to be confused with the ABC model of behavior (based on the works of Thorndike and Skinner), conveys that an attitude has three components: A for affective, B for behavioral, and C for cognitive. Although every attitude has these three components, any attitude can be based on one component more than another.

The Affective component is the emotional or feeling segment of an attitude. It is related to what affects another person. It deals with feelings or emotions that are brought to the surface about something, such as fear or happiness. Someone might have the attitude that they are happy with their job because they feel supported and enjoy their co-workers, or that they hate their job because it is frustrating and stressful to their health.

The Behavior component consists of a person's intention to act in a certain way towards someone or something. The behavioral attitude may be—"we better tell those guys to wear their bump caps" or "I'm going to try to get a promotion." The action of the individual is aligned with behaviors expected to reach a goal. The attitude focuses on the behavior.

The Cognitive component of an attitude refers to the beliefs, thoughts, and attributes that we would associate with an object (Schleicher, D. J., et. al., 2004). The cognitive is the opinion or belief segment of an attitude. It refers to that part of attitude which is related to the general knowledge of a person. Typically, these attitudes come to light in generalities or stereotypes, such as "all babies are cute".

In an organization, it helps to understand the complexity and the potential relationship between attitudes and behavior, especially when it comes to change. For example, imagine you realized that someone treated you with disrespect. You are likely to have feelings about that while instantaneously experiencing the realization? Thus, cognition and affect are intertwined in a complex relationship (Fabrigar, et. al., 2006). Attitudes are important because they often shape the perception of the individual, set the tone, and support the unconscious conditioning in a culture that integrates as part of the definition of expectations and requirements to succeed.

This aspect of the culture goes undefined and remains a mystery as part of the calculations of the effectiveness of a culture. Each one of these attitude components is quite different from the other, and they can build upon one another to form our overall attitudes and, therefore, affect how we relate to our world. Attitude is also our way of negotiating the gaps between self-determination and the expectations of others. This is an aspect of change management that rarely receives consideration and one that safety leaders need help with.

Early research suggested that attitudes were seemingly causally related to behavior (i.e., attitudes determine behavior). However, the ABC model of behavior would indicate that attitude is a consequence, which means it follows after the behavior is completed. Liu and Keng (2014) confirm the work of Festinger (1957) that attitudes follow behavior. Further research by Glasman and Ablarracín (2006) has shown that attitudes can predict behavior.

The fact that attitude follows behavior supports the effects of cognitive dissonance, which is described as the consistency between one's attitude and one's behavior, but when there are conflicts, between one's attitude and one's behavior, there is dissonance. Dissonance is when there is a clash (e.g., think of instruments in an orchestra warming up) resulting from the combination of two disharmonious or unsuitable elements, a lack of harmony in one's mind and heart. Dissonance is not healthy, and it's a noise in the mind of the individual that is dysfunctional and needs to be resolved in order for the individual to be fully functioning.

People seek consistency among their attitudes, and between their attitudes and their behavior. Any form of inconsistency is uncomfortable, and individuals will therefore attempt to reduce it. People seek a stable state, a minimum of dissonance. When there is dissonance, people will alter either their attitudes or their behavior, or worse, they will develop a rationalization for the discrepancy. Recent research found, for instance, that the attitudes of employees who had emotionally challenging work events improved after they talked about their experiences with co-workers (Edmondson, 1999).

The social phenomenon, regarding culture, operates when sharing with other co-workers their experiences, both positive and negative. This has helped workers adjust their attitudes towards behavioral expectations. No individual can avoid dissonance. You know texting while walking is unsafe, but you do it anyway and hope nothing bad happens (dissonance). Or you give someone advice you have trouble following yourself (dissonance). The desire to reduce dissonance depends on three factors: (1) the importance of the elements creating dissonance, (2) the degree of influence we believe we have over the elements, and (3) the rewards of dissonance; high rewards accompanying high dissonance tend to reduce tension inherent in the dissonance. In other words, dissonance is less distressing if accompanied by something good, such as a higher pay raise than expected. Individuals are more motivated to reduce dissonance when the attitudes are important or when they believe the dissonance is due to something they can control.

Psychological functioning

Our personality combines our attitude with our psychological functioning in a way that enables us the ability to harmonize our internal selves with the external environment. This harmony is what helps us fit in with an organization or a particular department because there is a good match between how we function internally and how the organization's leadership functions externally. As a leader, your ability to influence includes your individual behaviors, emotions, social skills, and overall mental health. Influence is grounded in having awareness and understanding of the cultural context and utilizing one's psychological functioning as a self-process for improving, developing, emerging, and ultimately performing at your best. You must lead yourself. Self-leadership depends on an individual's active interpretations of experience, with these interpretations affected, in part, by the meaning systems and practices emphasized in local cultural safety communities (Table 9.1).

Our personalities based on our psychological type function on a continuum of opposites. The two cognitive functions based on judging or perceiving consist of conflicting choices for behavior that pull from opposite directions, and each way of functioning activates a different part of the brain. When we make a choice to behave one way, it precludes the simultaneous use of the other way of functioning.

Thompson (1998) explains that when we employ a function, the continuum places us at a typological crossroads where we are axiomatically choosing against its opposite. Regarding our personalities we are forced to make choices in how we want to behave, and it results in each of us having to interact with the consequences of our choices and helps us evolve in our personal growth. Throughout our lives, certain kinds of adjustments are reinforced, and some are rewarded by the people around us, and others are

Table 9.1 Continuum of personality functions

Extroversion	*We can adjust ourselves* **or** *relate ourselves to an external situation*	*Introversion*
Sensation	***We can focus on what's in front of us*** *or* see other possibilities in our imagination	iNtuition
Thinking	***We can evaluate objectively*** *or* evaluate subjectively	Feeling
Judging	***We can organize events rationally and prepare for them*** *or* experience events directly as they happen	Perceiving

not. What gets reinforced has a tremendous influence on how we behave and what we are becoming as individuals. It also influences how teams behave (perform) in the workplace.

Descriptions of the four functions of personality

The structure of the four functions of personality is grouped into two categories: perceiving and judging. There are two functions in each category. The two perceiving functions are sensation (S) and intuition (N). Each of these functions is generally descriptive of encouraging us to keep our options and our minds open to gather information. These two functions are considered irrational because they are not predictable in that they happen with a regular pattern but are open to possibility. Sensation and intuition help us function in the way we take in information.

Sensing types function in the present moment of a given situation. They often express appreciation for all the particulars of the circumstances they are experiencing whether good or bad. Their intent is to gather literal information and seek to use that information for a utilitarian purpose. They spend a great amount of time gathering facts that are minute or complicated in their pursuit of understanding. They rely on past experiences when they solve problems and do not want to "reinvent the wheel." Because their purpose utilizes what is useful or practical, their ability to believe or trust something requires a reference to an experience or a substantive number of facts to determine its truth or value. They have excellent observation skills of their surroundings in how things smell, taste, feel, sound, or look and use that information for a precise execution towards a pragmatic outcome. They derive a sense of satisfaction from using things as they were meant to be used.

Intuitive types function best with the whole picture in mind. They recognize underlying patterns that aid them in creating a vision of future possibilities. Intuitives have a vivid imagination and seek to understand how patterns will lead to something in the near and far future. They are captivated by the unknown, the hidden, and the unseen. They have an appreciation for originality but also for alternative meanings. They rely on ingenuity during problem-solving and like to "reinvent the wheel." They are motivated when something new is a possibility. They hesitate to take on new responsibilities, relationships, or even purchase something because they feel they may be entrapped.

The other category is judging. The two judging functions are thinking and feeling. The use of either of these functions prompts us to understand how things actually happen and to organize our behaviors accordingly. Thinking and feeling are considered rational functions because they are based on predictability and that they happen regularly and in the same way. Thinking and feeling are how we make decisions based on the information we processed through our perceiving functions.

Thinking types make decisions based on logic. They analyze information objectively and impersonally. They have a keen interest in how things work, causal relationships, and general laws that apply to a process. Thinking types are bound to justice and ethical practices of fair play and equality. They are committed to individual freedom intellectually and consciously. They can anticipate obstacles, barriers, and objections to their plans. They like a logical, step-by-step approach to a conclusion.

On the other side of the continuum is feeling. Feeling types make decisions on a personal level, based on shared values and in the context of the relationship. Their interest is in how people feel about something—the emotional impact. They rely on consensus and have boundaries around morality, mercy, and loyalty. They are committed to social obligations and use empathy towards other people's situations. Feelers are natural empathizers, but they are not easily fooled by sad stories. Feelers are ultra-sensitive to fake emotions. They feel responsible to help others and anticipate people's needs, and how they will react. They are interested in relationships and have strong values by which they gauge and measure relationships. Feelers are good at picking up body language signals, and vocal intonations of how something was said to interpret how someone is feeling.

Carl Jung (1971 [1921]) discussed cognitive processing as part of the psychological functioning part of our personality. Jung described human beings as having two types of cognitive processes labeled perception and judgment. Perception involves all the ways of becoming aware of things, people, happenings, or ideas. Judgment involves all the ways of coming to conclusions about what has been perceived. There are two kinds of perception, sensation and intuition, and two kinds of judgment, thinking and feeling. Jung further explains that every mental act consists of using at least one of these four cognitive processes (i.e., sensation, intuition, thinking, feeling) by using one of the cognitive processes in either the extraverted or introverted world:

Extraverted (e) sensing (S) types and introverted (i) sensing (S) types

Extraverted sensing (Se) types are aware of the world around them in rich detail. They can determine factual data from general data in the confines of context and focus within that context to determine what immediate actions need to be taken to be successful. They actively seek a variety of inputs to understand the whole picture and desire an immediate result. They follow physical impulses or instincts that emerge from a situation and are excited to act in the present moment. They are in tune with the world around them and can become totally absorbed in what they are doing as they touch, feel, taste, hear, or see their experience. Se types are keen to visual input regarding their personal appearance, but also general appearances. They are quick to pick up on cues in a situation and like to push boundaries to get the desired impact they want and seek pleasure in both physical beauty and sensory novelty. They are commonly described as thrill-seekers or hedonists. The Se typically seeks work that involves physical interaction like an athlete, first responder, chef, artist, or mechanic.

Introverted sensing (Si) types have the ability to store and remember data or information that they can compare to their current situation with past situations. Their immediate experience or specific words link with prior experience to determine similarity or difference, like noticing a dessert is sweeter or too tart than it usually is. This tendency includes seeing people who remind them of someone else, history, experiences, and hindsight. The Si give great attention to detail to gain a clear picture of what success looks like regarding their goal, and what they want to accomplish. They have a oneness with traditions and customs that preserve culture, and they seek to protect what has been learned in the past. They like to preserve the past even while what is reliable changes. They can tune into their immediate inner sensations like when driving a special car and connect with its past and appreciate the journey it took for other people to create that car. The Si type cherishes the routine and familiar; however, they have to keep trying something over and over as they come to enjoy it more and more. The Si does not appreciate novel sensations as they are assessed as having less value or lacking authenticity.

Extraverted (e) iNtuiting* (N) types and introverted (i) iNtuiting (N) types

**N = intuition so as not to confuse it with introvert*

Extraverted iNtuitive (Ne) types have the ability to look at large amounts of data on spreadsheets or PowerPoints, information gathered from live interactions, videos or lectures, or behaviors based on their observations, and from all these information points, they can see patterns emerge and hidden potential meanings. They go beyond sense data (see, taste, touch, hear, and smell) of the Se or Si type. The Ne type discerns otherwise hidden patterns, possibilities, and potentials in the information. The information is used to create a strategy, or a concept often emerges in their minds. The Ne type utilizes brainstorming, drawing models on a white or glass board, and enjoys imaginative play with scenarios and combining possibilities with cross-contextual thinking using a variety of tools. The Ne type also can catalyze people and extemporaneously shape situations, spreading an atmosphere of change through emergent leadership. The Ne type enjoys reading, interesting conversations with authentic or educated people, and engaging with nature or the arts.

Introverted iNtuitive (Ni) types enjoy engaging with ideas, perspectives, theories, visions, stories, symbols, and metaphors to synthesize seemingly paradoxical or contradictory information and provide meaningful insights into a subject. The Ni cognitive processing occurs outside of their conscious awareness and therefore their best thinking is done without consciously thinking. The Ni type can have moments when completely new, unimagined realizations come to them. These realizations provide a picture of the future and a sense of urgency to act and stay focused on fulfilling the vision or dream of how things will be in the future. The Ni type will rely on graphic elements or visuals to predict, enlighten, or transform information or understanding that captures the vision. However, the Ni uses an emergent process of how a project will unfold, as unseen trends appear and telling signs reveal themselves—they point the way or bring together the pieces of the puzzle. This emergent process can involve working out complex concepts or systems of thinking or conceiving symbolic or novel ways to understand things that are universal. It can lead to creating transcendent experiences and solutions. The Ni type arrives at understanding through a single flash of insight—an "aha!" moment. This may occur while day or night dreaming with a surge of energy that keeps them moving on the journey of enlightenment.

Extraverted (e) thinking (T) types and introverted (i) thinking (T) types

Extraverted thinking (Te) types seek to make the external world more rational, to align how things work they like to plan and offer contingency plans, schedule, establish precise definitions, scripts, policies, and procedures that can be followed, and quantify data to organize their environment. To better understand and express their ideas, they use charts, tables, graphs, flow charts, outlines, and any creative visuals to clearly communicate the concept or information. The Te is sophisticated in their ability to organize work and observe people to be more efficient and productive, but to be optimized requires objectively to understand and control behaviors or conditions with standard operating procedures. These standards must be clearly explicated to minimize ambiguity and the potential for interpretative error. Towards this end, Te types often end up managing businesses or organizations. They are strong empirical thinkers and will challenge someone's ideas based on the logic of the facts in front of us or lay out reasonable explanations for decisions or conclusions made, often trying to establish order in someone else's thought process, or notice when something is missing like when someone writes there are four aspects of motivation, but only list three.

The Te is characteristically task or systems focused and comes across as impersonal. They literally think (i.e., ask questions, make factual statements, make logical judgments, conclusions, and decisions) out loud. They have a direct, "to the point" style of communication and sometimes are perceived by others as harsh, blunt, or tactless. They also compartmentalize many aspects of their lives so we can do what is necessary to accomplish our objectives.

Introverted thinking (Ti) types express a subjective logic with implied meanings. Instead of submitting to objective standards, the Ti reason and function based on inner criteria. They ask questions concerning the underlying premises and the assumptions that are being made regarding a concept, best practice, or model. They exercise an internal criterion of the essential qualities of something, noticing the fine distinctions that make it what it is and then they can label it, which is their way of coming to an understanding. The Ti's internal reasoning process derives the underpinning of subcategories of general categories and sub-principles of general principles. They build a schematic or framework on which key insights are understood. They are careful in finding just the right word to clearly express an idea concisely and to the point. These clearly expressed words can then be used in problem-solving, analysis, and refining of a task. This process is demonstrated in behaviors like taking things or ideas apart to figure out how they work by asking tough or critical questions. The analysis involves looking at different sides of an issue and seeing where there is inconsistency and comes across as criticizing or circumventing progress.

The Te-Ti have a natural tension that is regularly experienced in business relationships. The Te's function is always looking for ways to improve operations, which often involves implementing new policies and procedures. The Ti's function resists change and wants to keep things open ended until sufficient proof can be provided. The Te aligns itself with scientific method or objective methods. The Ti prefers a less formal approach and more on the individual level. This is the age-old distinction between the scientist (Te) and the philosopher (Ti).

Extraverted (e) feeling (F) types and introverted (i) feeling (F) types

Extraverted feeling (Fe) types tend to wear their emotions on their sleeves, and you rarely need to guess what they are feeling because their feelings are discernible through their mannerisms and facial expressions. Their desire to connect with others is evidenced by expressions of warmth and self-disclosure. They also seek to disconnect from others who are not genuine and will express displeasure in someone if they are not meeting their standards. The Fe function with social graces, and they are polite, nice, friendly, considerate, and appropriate. They keep in touch with people, laugh at jokes when others laugh to make others feel accepted or comfortable, and like getting people to get along with others. Fe types are looking to create a bond of shared feeling, especially a good feeling. They readily express their own feelings while perceiving and interpreting the feelings of others. They desire that feelings will be understood and reciprocated in a way that allows all parties to enjoy a sense of emotional resonance and harmony. The satisfaction of creating rapport and emotional harmony also leads Fes to enjoy supporting and counseling others with emotional or relational difficulties.

The Fe's function responds according to expressed or even unexpressed wants and needs of others. They may ask people what they want or need or self-disclose to prompt them to talk more about themselves. The Fe engage in conversation to learn more about others so they can better adjust their behavior to the other person. This makes the Fe feel responsible and take care of others' feelings, sometimes to the point of not separating our feelings from theirs. They recognize and adhere to shared values, feelings, and social norms to get along. The Fe have broad and extensive feelings for the collective morale of the group and tend to prefer external harmony because they are personally not comfortable with conflict. They are more inclined to help others manage their emotions, as well as to turn to others for emotional support and to get everyone on the same emotional wavelength. Fe's expressions are more direct and feeling-laden, conferring a sense of greater urgency or conviction in what they are saying. At times, it can feel like Fe's (especially iNtuive, Feeling, Judger (NFJ) types NFJs) have fallen into a motivational speech or diatribe during an ordinary conversation.

Introverted feeling (Fi) types see themselves as managers of their own emotions, and they want others to do the same. The Fi are more emotionally independent and tend to restrain or conceal their emotions. They present a more outwardly measured persona and are less animated in their gestures and expressions. Introverted feeling (Fi) is one of the hardest cognitive functions to understand, explain, and observe because it is an introverted function. It is taking place inside the mind where others cannot see it.

The Fi type is often associated with images, feeling tones, and gut reactions more than words, and their cognitive process serves as a filter for information that matches what is valued, wanted, or worth believing in. They penetrate more deeply and intensively while focusing on the feelings of the self or a select few individuals. There can be a continual weighing of the situational worth or importance of everything and a patient balancing of the core issues of peace and conflict in life's situations. The Fi is the most sensitive of all types. They can react emotionally for no obvious reason. It takes some effort to drill down and uncover these personal values that are so easily insulted. They are sensitive when a value is compromised and will speak up to clarify their feelings. Much of the emotional processing is internal but expressed through actions. They seek to determine if people are being fake or insincere, or if they are good. It is like having an internal sense of the "essence" of a person or a project and reading fine distinctions among feeling tones. The Fi adapt to circumstances and situations, both external and internal, and they learn by experience and preference to react and to respond using our cognitive functions. As their functions become more developed, they become more aware of their natural skill sets, and those certain preferences are easier to utilize. Interestingly, despite being a feeling function, Fi is not really a social (or socializing) function. Fe is characteristically *interpersonal.* Fi, by contrast, is *intrapersonal.* It involves a relationship with oneself, with one's own emotions, tastes, and values.

When you understand how you function, you will realize that you have natural strengths and weaknesses. This is where self-influence becomes a stronger possibility. Some people think it is logical to work on their weaknesses because if you can improve those and add them to your strengths, you will become a whole person of strengths. Logic is not science, and in this case, it is not supported in personality theory. With personality traits, a strength and a weakness are on opposite ends of a continuum. That means every weakness has a strength, and every strength has a weakness. This also means that if you improve a weakness, you also diminish the strength. This also works the same way if you improve a strength, you then diminish the weakness. What this means is that we can never get rid of our weaknesses or our strengths—we will always have them, and perfection is not possible. This logic then, seeking to improve a weakness, is a waste of time regarding your personality. People often confuse improving weak knowledge, ability, attitude, skills, or industry-specific insights (KAASISIs) with personality, but they are two different things. You can always improve your KAASISIs, but my recommendation, regarding your personality traits, is that you focus on improving your strengths. Your awareness of your weaknesses will keep you humble and grow your appreciation for your strengths and the personality trait strengths of others. This is a critical understanding of the development of leadership. You must know your personality strengths and your limitations and most importantly focus on developing your strengths. When it comes to personality strengths and weaknesses, there is no need to fix your weaknesses—you will always have them. This will help you as a leader to improve your self-leadership but will also help you to influence other individuals. One-on-one influence is the foundational level of leadership. Beyond this level is the advanced level, the ability to influence groups. Beyond the advanced level is the competent level where the ability to influence conditions within a culture is the ultimate ability to influence mass amounts of people.

LEADERSHIP STYLES

Kotter (2001) stipulates that a leadership style is a leader's method of providing direction, implementing plans, and motivating people. Research has identified many different leadership styles as exhibited by leaders in politics, business, and other fields. An effective leader influences followers in a desired manner to achieve desired goals. The greater the influence, the greater the power to achieve desired goals. Different leadership styles may affect organizational effectiveness or performance. Organizational performance is influenced by a competitive and innovative culture. Organizational culture is influenced by leadership style, and consequently, leadership style affects organizational performance. In the world of safety there needs to be less transactional leadership and more transformational and servant leadership.

Transactional (authoritarian) leadership

Demanding to be in control of everything is the motivation for telling people what to do and a significantly damaging leadership habit that severely diminishes safety performance. When I conduct an onsite visit, I observe the safety team in action, listen to the leaders, experience their environment doing walk arounds, listen to people explaining their work and their understanding of their role and job description, read their policies and procedures, review metrics and data reporting process as well as the technology, and participate in a variety of safety meetings, and I keep coming across one common thread that seems prevalent everywhere in safety and that is the practice of *telling* people what to do.

Telling people what to do is a strong habit of safety leaders (and business leaders as well—maybe worse). Telling people what to do is part of the transactional leadership style that supports a mechanistic culture. The traditional safety leader is a transactional leader who lacks empathy and follows the letter of the law and not the spirit of the law. Transactional leadership focuses on results only, conforms to the existing structure of an organization, and measures success according to that organization's system of rewards and penalties. Transactional leaders have formal authority and responsibility in the organization and often become authoritarian leaders. Leadership is influence. However, not every person has the ability, capacity, or capability to influence at the individual level, group level, organizational level, or industry level. The lack of influence drives a person to leverage their authority. This type of leader maintains low levels of performance, which induces stress, pressure from executives, and personal pressure to perform.

This type of leader sets the criteria for their workers according to previously defined requirements (doing what they have always done). Transactional leaders continually up the requirements or raise the bar, so to speak, to "get more" out of their people. The consistent demand is more, more, and more! An annual performance review is a brutal tool used to punish people. The performance review is usually based on the last six to eight weeks of performance, and the previous ten months are ignored because transactional leaders live by the mantra "what have you done for me lately." Transactional leaders work best with employees who know their jobs and are motivated by the reward-penalty system and need people who are willing to play their game, and it's a brutal game of survival. Employees are greatly rewarded and severely punished. The roller coaster of emotions, lack of empathy, and zero psychological safety burns employees out at a high rate. The status quo of the organization is maintained, and this frustrates the transactional leader who becomes more and more dictatorial, blaming people for not reaching their goals, for not responding fast enough, blamed for anything the dictator wants as long as the leader is not blamed or held accountable for their failed ability to influence. This is not BBS, this is RBS, and yet BBS is blamed.

Telling people what to do has a cascading waterfall effect—the business leader tells the safety leader what to do, and the safety leader then tells the safety manager what to do, and then the safety manager tells the safety supervisor what to do, and the safety supervisor tells the safety team leader what to do, and then the safety team leader tells his or her team what to do. It gets so bad that safety team members tell each other what to do. They go home to their families and tell their kids what to do, and the kids tell the dog what to do. The poor dog is so confused! The way a leader uses his position impacts the attitudes of the employees. *Telling* people what to do squelches their psychological safety and drives fear into the hearts of people.

Psychological safety is being able to show and employ oneself without fear of negative consequences of self-respect, image, status, or career (Kahn 1990, p. 708). It can be defined as a shared belief that the team is safe for interpersonal risk taking. My research has revealed that *telling* people what to do is a significant contributor to dysfunctional behavior, and it damages the safety culture. The waterfall of telling falls at such a rate of speed that the last person at the bottom gets pounded down so hard that it adversely affects their performance. It is terrible to do that to people. No one likes to be told what to do. They tolerate it because of the hierarchy (fear), but they do not like it or enjoy it, and it shows up in the performance metrics (if you have them).

Psychological safety does not take root when a lack of empathy exists. Transactional leaders never use empathy. However, I've learned that the use of empathy is significant, and yet 99% of people use it incorrectly. Most people make an apology "I'm sorry" and call that empathy. Few people think about what they are saying, how their words are used, and the impact of those words. Effective communication is using the right words accurately. How is an apology empathy? It's not. Empathy is the ability to share the feelings of another. Expressing empathy properly is to share the feelings of the other person. Instead of apologizing, try responding with a heartfelt statement. For example, someone shares that they are frustrated working with a co-worker. An empathetic response would be, "well, frustration is no fun, what's happening to cause you frustration, and how can I help you?" If an employee is struggling to get a task done, you could engage in a conversation by saying, "Hey, it looks like you're struggling, could

I be of any assistance?" Empathy is about sharing the feeling of another. Using empathy starts by reflecting back to the person what you see in their face or what you hear in their voice. Someone looks stressed, so you say, "you looked stressed, do you want to talk about it?" Someone comes up to you and is stressing out, "I'm listening to you, and you sound stressed, take a breath, why don't we sit down, and I'll continue to listen and do my best to help you." It's important to express empathy correctly because it builds healthy relationships and earns trust. People trust people who are empathetic.

Transactional leaders rarely earn the trust of their people because their leadership style instills fear. Transactional leaders lead mechanistic cultures and breed an attitude of elitism that promotes a condescending tone of the conversation from leadership down to the people. This is not helpful, but it goes on in so many safety programs that exist in a hierarchical style organization. Transactional leaders in their mechanistic cultures pursue Results-Based Safety. What you practice is what you believe. It seems in the world of safety that telling people what to do and being seen as the ultimate source of knowledge, being politically connected, and acting as if they are safe in their job regardless of what they do is a prevalent theme I have seen over and over. It is a true blind spot for safety leaders to flaunt the fact that their ignorance, lack of performance, and lack of leadership will not get them fired, and is more valued than the safety of the people. This needs to change immediately. Business and safety leaders need to consider the serious repercussions of these conditions in their culture. It undermines the safety program, and that makes business and safety leaders responsible for what happens. This includes legal action. Why do executives tolerate these risks? Because all they want are results.

Business and safety leaders need to stop acting as if they understand the science of human behavior because they attended a workshop. They need to be held accountable for proper practice of the science of human behavior (i.e., BBS, Human Factors, Human Analytics, and Anthropology) just like any other aspect of the business. These are not buzz words, and many safety leaders mesh these concepts together and the result is confusion, chaos, and misinformation. Safety leaders may think they look smart, but they are not, and their performance outcomes are not there. They may be achieving their safety results, but I doubt very much they can explain how they achieved those results. Business and safety leaders need to be transparent about how they measure and produce the numbers they present. It should not be a secret practice in a dark back room, and surprise, we got our results! When people ask, "How did we get our results?" The answer from leadership should not be, "It is complicated, and you do not need to worry about that."

Value the ability, to ask people what they think, not *telling* people what to do. I find this issue of *telling people what to do* to be the literal cancer that undermines almost every safety program out there because it creates a toxic and dysfunctional mechanistic culture. My research shows that *telling* people what to do has an adverse impact on safety behavior and ultimately on safety performance because it hurts people, and it destroys relationships. *Telling* people what to do, even if you do it in the nicest way possible, most times comes across as condescending. It comes across as "I know more than you," "I am in charge," "you're not important," "I don't care what you think," "it doesn't matter what you think," and "I don't believe that you think"; these are the messages often communicated non-verbally, and none of them helps a person to be better or challenges a person to improve how to think more safely.

My research shows that when you tell people what to do, they will comply in the immediate circumstance, but they usually do not comply later when the leader is not there. This is not a good safety behavior—do what the boss says when he or she is there but ignores them when they are not there. This risky behavior is reinforced by the safety leader walking around and telling people what to do. *Telling* is a corrosive behavior, habit, and value that literally undermines the possibility of having healthy functional relationships. *Telling* is a habit, and it happens everywhere because it is part of the culture. Here is an example of *telling* in a typical safety communication.

Example of a safety leader communication

Welcome back, hope you guys had a happy new year! Just a couple of safety tips and reminders to start the year off.

- Most kids go back to school on Monday, January 7th which means:
  - School zones
  - Kids walking to and waiting at the bus stops

  If you must drive in foggy conditions, keep the following safety tips in mind:
- Slow down and allow extra time to reach your destination.
- Make your vehicle visible to others both ahead of you and behind you by using your low-beam headlights since this means your taillights will also be on. Use fog lights if you have them.
- Never use your high-beam lights. Using high-beam lights causes glare, making it more difficult for you to see what's ahead of you on the road.
- Leave plenty of distance between you and the vehicle in front of you to account for sudden stops or changes in the traffic pattern.
- To ensure you are staying in the proper lane, follow the lines on the road with your eyes.
- In extremely dense fog where visibility is near zero, the best course of action is to first turn on your hazard lights, then simply pull into a safe location such as a parking lot of a local business and stop.

- If there is no parking lot or driveway to pull into, pull your vehicle off to the side of the road as far as possible. Once you come to a stop, turn off all lights except your hazard flashing lights, set the emergency brake, and take your foot off of the brake pedal to be sure the tail-lights are not illuminated so that other drivers don't mistakenly run into you.

Example of a safety leader communication [with translation of how it comes across]
Welcome back, hope you guys had a happy new year! Just a couple of safety tips and reminders *[reminders are a kind way of "telling" people what to do]* to start the year off.

- Most kids go back to school on Monday, January 7th which means:
  - School zones
  - Kids walking to and waiting at the bus stops

  If you must drive in foggy conditions, keep the following safety tips in mind:
- *Slow down* and *allow extra time* to reach your destination. *[this is a "tell"]*
- *Make* your vehicle visible to others both ahead of you and behind you by using your low-beam headlights since this means your taillights will also be on. *[tell]* Use fog lights if you have them. *[tell]*
- *Never [dramatic emphasis "tell"]* use your high-beam lights. Using high-beam lights causes glare, making it more difficult for you to see what's ahead of you on the road. *[potentially condescending explanation on how high-beam lights work based on context of telling]*
- *Leave* plenty of distance between you and the vehicle in front of you to account for sudden stops or changes in the traffic pattern. *["tell"]*
- To ensure you are staying in the proper lane, *follow* the lines on the road with your eyes. *["tell"]*
- In extremely dense fog where visibility is near zero, the best course of action is to first *turn on* your hazard lights, then simply *pull* into a safe location such as a parking lot of a local business and *stop. ["tell," "tell," "tell"]*
- If there is no parking lot or driveway to pull into, *pull* your vehicle off to the side of the road as far as possible. Once you come to a stop, *turn off* all lights except your hazard flashing lights, *set* the emergency brake, and *take* your foot off of the brake pedal to be sure the tail-lights are not illuminated so that other drivers don't mistakenly run into you. *["tell," "tell," "tell"]*

The leader in this communication uses a tone (whether it is intentional or not) that comes across as an expert in driving, school zones, kids, parenting, driving kids, driving as a parent, school zone law, vehicle visibility, driving safety, headlight candescence, headlight visibility, high-beam projection, fog light usage, visibility in general, vision, eyesight clarity, vehicle distancing, stopping, dense fog conditions, parking, off-road parking, etc. Nobody appreciates being talked down too. Most people are left feeling fear, dread, stupidity, and seriously doubt their capability to make a simple trip to school to drop their kids off—all based on this safety leader's authority, position, and attitude. Wow! This is good doom and gloom communication—making people feel less capable and focused on getting results. Let's look at an accurate BBS sample.

Example of a safety leader communication (an effort to get people to think safely)

> Welcome back and I hope you all had a happy new year! Just a couple of safety tips and reminders to start the year off.
>
> - Most kids go back to school on Monday, January 7th which means:
>   - School zones
>   - Kids walking to and waiting at the bus stops
>
>   *When you* drive in foggy conditions, here are some questions to ask yourself to think about driving safely:
> - What is an appropriate speed to drive safely in foggy conditions?
> - How much time should you allow yourself to reach your destination safely?
> - How could you make your vehicle more visible to others both ahead of you and behind you?
> - What impact would using your high-beam lights have on the safety of other drivers?
> - What is a safe distance between you and the vehicle in front of you to account for sudden stops or changes in the traffic pattern?
> - How can you ensure you are staying in the proper lane?
> - In extremely dense fog where visibility is near zero, what is your safest plan of action?
> - If there is no parking lot or driveway to pull into, what is your plan of action?
>
>   I hope you find some of these questions helpful as you plan a safe and enjoyable trip to school. Happy New Year!

This style of communication helps people focus on safety, it reinforces their capability to think and make accurate decisions based on awareness and being present with current conditions. This style of communication believes in people and trusts people. Communication is an antecedent that prompts or reminds a person to act in a safe way. The consequence of positive reinforcement is a strong possibility here—that the parent will drop their child off safely and return home or arrive at work safely. The focus is on the behavior, not the result. The leader comes across as supportive, not commanding, not an expert, but helpful and concerned about the safety of all those people traveling in winter conditions and arriving at school safely, but also returning home safely. People can sense that concern. It goes a long way with them, and they appreciate it when leaders are truly helpful.

Creating an effective safety culture is an exceedingly difficult challenge and it requires measuring observable desired and required safe behavior. Creating functional and transparent communications that are aligned with the corporate culture is healthy. It's the only true option for achieving sustained safety results. A safety culture is a difficult work, and there are a lot of complexities; however, it is achievable and measurable, but it takes effective leadership. An effective leadership style would be transformational leadership.

Transformational leadership

Transformational leadership is a strong predictor of both job satisfaction and overall satisfaction. James MacGregor Burns (1978) first introduced the concept of transforming leadership in his descriptive research on political leaders, but this term is now used in organizational psychology as well. Transformational leadership is defined as a leadership approach that causes a change in individuals and social systems. In its ideal form, it creates valuable and positive change in the followers with the end goal of developing followers into leaders. Enacted in its authentic form, transformational leadership enhances the motivation, morale, and performance of followers through a variety of mechanisms. These include connecting the follower's sense of identity and self to the vision, purpose, values, and collective identity of the organization; being a role model for followers that inspires them; challenging followers to take greater ownership of their work, and understanding the strengths and weaknesses of followers, so the leader can align followers with tasks that optimize their performance (Schuh et al. 2013). The use of empathy creates psychological safety, and this is what builds trust in the safety leader.

Bass (1990) explains that transformational leaders provide individual focus and motivation by challenging their people to reach their goals. Transformational leaders motivate and teach with a shared vision of the future. They communicate well. They inspire their group because they expect the best from everyone and hold themselves accountable as well. Transformational leaders usually exhibit the following character traits: integrity, self-awareness, authenticity, and empathy. Another effective leadership style is servant leadership.

Servant leadership

Servant Leadership is a fundamental leadership style that is sorely needed in the world of safety. Greenleaf (1970) defined servant leadership as a philosophy and set of practices that enrich the lives of individuals, build better organizations, and ultimately create a more just and caring culture. A servant leader focuses primarily on the growth and well-being of people and the communities to which they belong. Based on the research of Spears (2010), the ten most important characteristics of servant leaders are listening, empathy, healing, awareness, persuasion, conceptualization, foresight, stewardship, commitment to the growth of people, and building community. Safety leaders would do well to adopt to a servant leadership style to be more effective in a BBS safety program.

Servant leadership is different from traditional leadership where the leader's main focus is the thriving of their company or organization. Sendjaya and Sarros (2002) explain that a servant leader shares power, puts the needs of the employees first, and helps people develop and perform as highly as possible. Servant leadership inverts the norm, which puts the employees as the main priority. Instead of the people working to serve the leader, the leader exists to serve the people. As stated by its founder, Robert K. Greenleaf, a servant leader should be focused on, do those served grow as persons? Do they, while being served, become healthier, wiser, freer, more autonomous, and more likely themselves to become servant leaders? When leaders shift their mindset and serve first, they benefit as well as their employees in that their employees acquire personal growth. While the organization grows so do the employees in their commitment and engagement. Since this leadership style came about, a number of different organizations have adopted this style as their way of leadership successfully.

Sendjaya and Sarros (2002) claim servant leadership is being practiced in some of the top-ranking companies, and these companies are highly ranked because of their leadership style and the following it creates. Further research also confirms that servant leaders lead others to go beyond the call of duty. In psychology this is called discretionary effort. Since the turn of the century, servant leadership has been popularized in software development through the Scrum and Agile management methodologies. A servant leader is someone, regardless of their level on the corporate hierarchy, who leads by meeting the needs of the team (Greenleaf, 1970). Values are important to the servant leader, and those who lead within this leadership style do so with the generosity of spirit. Servant leaders can achieve influence and power because their ideals, ethics, and use of empathy resonate with their people, and they earn the trust of their people. Safety leaders can adopt a leadership style, but with it they have to practice, that is demonstrate the basics that make leaders effective.

LEADERSHIP PRACTICES

Vision

Vision is where you want to go. It is very frustrating to see and experience the number of safety programs that are based on what *not* to do (RBS). That is not a vision, that's a reverse vision, where you don't want to go. You cannot create a healthy culture that produces success by establishing a vision of what *not* to do. The faulty logic here is based on the idea that if people do not do a certain behavior, it will automatically be replaced with the desired or required behavior. This is incorrect. Behavior does not work this way.

When it comes to vision, business and safety leaders must establish an inspiration of what can be. A vision must inspire hope, confidence, and a direction that establishes conditions that are a true possibility. A successful vision is not just about achieving safety numbers, that is Results-Based Safety. A vision is about the possibility. The ultimate safety vision is to establish a culture that reinforces safety thinking and focuses on desired safety behavior—that is Behavior-Based Safety. Remember, results are an indicator that you are reinforcing the right behaviors frequently and fluently. Beyond the vision, a safety leader needs a clearly defined purpose.

Purpose

Purpose is the why of what you are doing. Establishing a purpose for your safety program is probably the most misunderstood, and most damaging aspect to safety efforts. This is where the misunderstanding of Results-Based Safety (RBS) and Behavior-Based Safety (BBS) clashes. Business leaders tend to be concerned with numbers or results because they are incentivized (rewarded) to achieve those results. However, let us get some clarity here right up front. Safety is about people, their behavior, and the culture that creates the conditions that they work in to think safely and behave safely. The results are an indicator that you understand your people, reinforce safety thinking and behavior frequently and fluently, and are continuously monitoring your culture properly.

A business or safety leader had better understand this purpose quickly, or they need to move on to another position. Accountability starts with the top business leader and the top safety leader. There is no place for Results-Based Safety in safety. There should be no incentive, no bonus, and there should be nothing that benefits a business leader or any safety leaders for achieving safety numbers.

The true purpose of any safety program should be to establish a safety culture that reinforces safe thinking and safe behavior. Reinforcing the thinking and the behavior is the key, and if you are doing it right, you will get your results. It seems that the majority of the controversy about BBS is based here. The confusion is based on clarifying and distinguishing RBS from BBS. These two methods are in direct opposition to each other. People keep bashing BBS and accusing the methodology falsely because of RBS-related tactics and practices. "Blame the worker" is RBS. The majority of controversy and criticism around BBS is because people keep mistaking it for RBS. Results should be viewed as an indicator that you have accurately aligned your culture with clearly defined safe behaviors, established leading indicator metrics, practiced performance coaching, and achieved the required outcomes of the safety program through a supportive system where people work together. Attaining results should not be the goal. My research has shown that when you increase the number of reinforcements, then the number of desired and required safe behaviors also increase, and the correlation is significant. The more you reinforce desired and required safe behavior, the more the defined safe behavior occurs and drives those safe behaviors to habit strength.

The real goal in safety is to get 100% of your people to think and behave safely, not zero results today (RBS). Safety leaders should be posting the number of safe behaviors observed daily, the number of positive and negative reinforcements daily. Reinforce people to reinforce other people. How about posting stories that capture safe thinking, safe planning, and safe solutions that have been discovered, enacted, and worked successfully. BBS is about people doing safety, safely.

Safety leaders must reinforce the desired and required behavior by reinforcing safe thinking. When everybody thinks and then behaves safely, the near misses, incidents, and injuries decrease. And if you follow the seven steps of Performance Safety Coaching©, you will dramatically improve engagement in how people interact with each other and behave safely by improving relationships. When people are empathetic towards each other, and psychological safety is present, trust is earned. It is important that safety leaders are careful to adopt an attitude that we are all in this together. Safety is a collaborative effort. RBS develops an attitude that separates safety leaders from the front line or field people. Safety leaders often say one thing but practice another. If you observe their behaviors and listen to the way they talk to people, it becomes obvious that results are their true concern. One major indicator is the practice of *telling* people what to do, and this practice is a strong indicator that the safety program is dysfunctional and will not be effective or meet its required objectives.

Business leaders and safety leaders need to come together and learn the seven steps of Performance Safety Coaching©. Learn how to reinforce one another and model for everyone else how conversations are to happen, and model how to treat one another. No more hiding in the office, no more "let the people in the field" focus on reinforcements, or let the lower levels focus on reinforcements. When business leaders act as if safety is for the little people, it completely undermines and devastates the safety program, and it is a sign that RBS is being practiced not BBS. Business leaders can do better and need to do better. Safety leaders need to be reinforced on how to act because it is what reinforces them to behave the way they do.

The focus of safety leaders should be on creating conditions that support safety thinking and reinforce desired and required safe behaviors. The ultimate goal would be for them and their people to become fluent in the seven steps of a Performance Safety Coaching© conversation and practice frequent reinforcement of desired safe behavior. But, before safety leaders can become fluent and frequent in their reinforcement skills, someone has to teach them how to do it, and they themselves have to be reinforced in the behaviors of safety coaching until they have achieved habit strength of the seven steps of Performance Safety Coaching©. Unfortunately, it is not natural, and Performance Safety Coaching© has to be learned, practiced, and practiced again until it becomes habit strength. Reinforcing behavior is an elusive skill set, and the skill set requires strong reinforcement of those leadership behaviors if safety leaders are going to be successful.

Clarity of purpose is essential to a successful leadership program. Get a clear vision and a clear purpose and focus on the right things, like reinforcing desired and required safe behavior. Reinforcing desired and required safe behavior is made clear by what you value. Clarified values are a necessity, but how you practice your values sends a message that is psychologically stronger than any other efforts you make. It is painfully obvious what really matters to leadership because it is demonstrated by how leaders practice what is valuable in the context of your culture.

Values

Values are how you do and what you do. Almost every organization has vision, purpose, and value statements. It is vitally important to an organization to define and practice these statements, and how they are specifically applied to safety. It is also important to everybody in the organization to see how leaders practice what they value, and it carries more impact on your people than anything else. How you practice your values is a direct revelation of how your culture functions. Business and safety leaders do not always realize the impact they have on the performance of their employees because their behavior is causally related to how they practice the organization's values.

If a business leader claims they value treating people with respect but lets a safety leader get away with treating others with disrespect, do not be surprised when your engagement scores are low in that leader's area, or performance struggles in that leader's area. What some business leaders truly value is the fact they do not care how you get something done, but that you get it done. Just get the results (RBS).

This mentality is not accepted currently by the younger generations of X, Y, and Z. Top leaders often work extremely hard at keeping it a secret from the majority of the organization how they think about people in their organization and yet it is observable in how they treat people in the organization. Top leaders hire speech writers and social media experts to portray a faux image of who they really are. Authenticity and transparency have disappeared. Top leaders charge HR leaders with keeping all these damaging issues clandestine, and the image and reputation of HR suffer for their compliance.

Talented young people are aware of how a healthy or dysfunctional organization functions, and they are not putting up with the dysfunctional ones, and it is costing these businesses dearly. Business leaders can no longer work in isolation from the front lines. Business leaders as well as the entire organization must practice what is valued by all not just what they personally value ($).

In the world of safety, it is no different; safety leaders must practice what is valued, not what they value. The two issues should be aligned. There is so much confusion on how to do safety because of the influence of business leaders who practice RBS. There are so many voices out there screaming for attention, demanding to be heard, and *telling* you what you need to do. Safety leaders have to develop their ability to influence.

LEADERSHIP INFLUENCE

Leadership is influence. Leadership has been referred to as a social influence process because the leader seeks the participation of employees in an effort to reach organizational goals. A leader delegates tasks and responsibilities to act so as to carry out specified objectives. Organizations need effective leaders who understand the complexities of the rapidly changing business environment, but they also need leaders who understand how to motivate the workforce and accomplish goals. If the task is highly structured and the leader has a good relationship with the leadership team and the employees, the effectiveness will be potentially high on the part of the employees. The pivotal issue for a leader is participation, and how that leader gets the participation of people. There are two major types of leadership styles: authoritarian and democratic.

When a leader cannot get people to participate through influence, they leverage their authority. The authoritarian leader lacks the knowledge, attitude, abilities, skills, and insights to motivate people. The transactional leader becomes more authoritarian and lacks the necessary relational skills and personal motivation to see others develop and perform at their best. The authoritarian leadership style is characterized by abuse of power, monopolizing decision-making, and being dismissive of alternative opinions. The transactional leader becomes selfish and narcissistic and suffers from serious cognitive dissonance. That is they know they could do it better but don't care. The pressure to get results is their focus.

A transformational or servant style leader engages with the members of the group, and the members take a more participative role in the decision-making process. Researchers have found that these styles are usually the most effective in supporting people to higher productivity, better contributions from group members, and increased group morale. In psychology this is called discretionary effort. Discretionary effort is emotional, the individual owns the work they are doing, and they will do what it takes to get the job done.

The transformational or servant style of leadership offers everyone the opportunity to participate, exchange ideas, and have their opinions heard, encourages discussions, and allows people to engage at a level to their satisfaction. Employees engage in their work at a higher level because they believe in the leader, the vision, purpose, and values of the organization and demonstrate discretionary effort. Discretionary effort occurs when employees are recognized for going above and beyond, striving for excellence, and keeping the interests of the company top of mind. Discretionary effort creates and reinforces a culture of strong employee engagement and is the highest order for a leader to achieve. A leader serious about expanding their abilities to influence large groups of people needs to understand and expertly influence conditions, relationships, and culture.

Influencing conditions

Influencing the conditions of the environment is what shapes large groups of people's behaviors. What are the conditions? Conditions are the vision, purpose, values, circumstances, situations, policies, procedures, attitudes of employees, leadership styles, office politics (good and bad), daily interactions, individual well-being, individual health, quality of relationships, and other intangibles that affect the way people live and work. For example, safety would be a condition that affects the way people experience life, and well-being is a condition that affects the way people experience life.

In the bigger picture there are psychological constructs at play, and these conditions can be categorized into four buckets: the mental, social, physical, and spiritual aspects of life that people experience. There is a lot to think about here regarding conditions. As a safety leader, your mind should be focusing right now on the conditions that your people experience every day at work. What are those conditions? Can you identify and define them? How do those conditions effect your people? How do those conditions effect your safety campaigns? Desired safe behavior? Safety performance? Safety outcomes? You will need to investigate and learn about the current conditions that exist in your safety world. Have discussions with your people, talk about it during safety meetings, talk to leadership and other safety professionals, and listen to all the feedback. You will need to take notes, organize your notes into categories, and identify and define every condition. Use your safety meeting to involve the team in identifying and defining current conditions, conditions that need to be stopped, conditions that need to change, and new conditions that need to be added to the safety culture.

The real secret to effective safety leadership is having a plan to create and influence the conditions of the safety culture in which everyone chooses to engage in thinking and behaving safely. This is the hard work of safety leaders and far too often they think it is easy or do not think about it at all. The ability to influence individuals is a critical skill, but influencing greater numbers of people is a crucial element of the safety leader's competency.

Unfortunately, safety leaders focus solely on influencing individuals or small groups of people, known as clicks and silos, which are extremely limited to a few people or a few groups of people. There are a lot of clicks and silos in safety, and that all adds up to group think. Group think is a psychological phenomenon that occurs within the social phenomenon of a culture. It occurs within a group of people in which the desire for harmony or conformity in the group is strong enough that it results in an irrational or dysfunctional decision-making outcome. What metrics do you keep around group think? As a safety leader, do you recognize when this is occurring in your culture?

Many safety leaders take the easy way by creating a mechanistic culture, which is to leverage their title and authority to *tell* people what to do, yell at people, punish or penalize people, and use scare tactics and intimidation to get results. These weak leaders have created a bunch of dysfunctional conditions (e.g., fear, fear of failure, feelings of inadequacy, and anxiety) that undermine their very efforts to create a sustainable culture. It is all about the numbers (RBS) but look at the numbers. The evidence shows that bad things still happen. Unsafe behaviors still occur and that is because dysfunctional conditions exist.

Identifying dysfunctional conditions

There are a few obvious dysfunctional conditions that business and safety leaders need to be aware of and change. Dysfunctional conditions germinate and grow in mechanistic cultures. Mechanistic cultures centralize their decision-making, create formal, standardized control systems, and establish stable, simple environments. Managers integrate the activities of clearly defined departments through formal channels and in formal meetings, and often robotically. Mechanistic cultures have many hierarchical layers and a focus on reporting relationships, centralized decision-making by a few people, clear hierarchy of who is in authority, with clear and efficient reporting relationships.

They are essentially bureaucracies. The benefit is a smooth operating machine that is predictable and sustained in its performance. The tendency is for the operation to run itself, leaders become insouciant, intransigent, dramatic, complacent, and ineffective in that they bottleneck communication, information, processes, and growth. The leader's focus is on themselves; they create a blind spot as a lid to the professional opportunities for growth, performance, and career development for their people. The struggle for mechanistic leadership is real, and here is what the struggle looks and feels like.

First, a weak safety leader cannot get people to do what they need done, so they heavily leverage their title. They communicate that they are in charge. That they have the responsibility and that they are connected with upper leadership.

Second, because they have a difficult time influencing people, they have a strong propensity to tell people what to do and it is not an effective way to influence people. Telling (RBS) is not coaching (BBS). Unfortunately, safety leaders think that telling people what to do works because it has an immediate impact, and it does to a point. The problem is that most of the impact that safety leaders have is a negative impact.

Third, they do not see that they have a negative impact, but it shows up in incidents and injuries. Weak safety leaders do not collect behavioral data (i.e., leading indicators) so they never see the correlation between telling people what to do and unsafe behavior(s). I have listened to lots of safety leaders brag about how they do not have any injuries. When a safety leader brags about zero injuries, it is obvious that their focus is on results. I like to ask the safety leader a few questions: could you explain exactly how you are achieving zero injuries? What is your method? What are your metrics? What specifically is your strategy? What leading indicators do you use? How do you implement consequences into your safety program? How have you defined desired and required safe behaviors?

Safety leaders get defensive at these questions. That indicates to me that they do not know how they are getting zero results (and the irony of not getting results), and that is not a good thing. Anytime a safety leader or business leader gets defensive and secretive about how they are getting their results, it usually means they are cheating, or they honestly do not know, and they should admit that they honestly do not know. It also means that they have created conditions in the work culture where workers may feel threatened to report any injuries. A powerful side effect of safety leaders telling people what to do is that their people do not feel comfortable to do the right thing. Because of these conditions, up and coming safety leaders practice telling others what to do. It is what they see and hear all day and every day, and they are reinforced to practice telling others what to do instead of how to think. Because they are told they're doing a good job (which is positive reinforcement) and, ultimately, they are promoted. This sends a strong message to everyone in that department on how to get ahead, but it is the wrong message. It is these types of conditions that shape how people behave in a limited capacity.

Telling people what to do and using one's authority instills fear into people to not question the leader because of their positional and political power. And if anyone even tries to question leadership, they are usually punished in some discreet way, and the message is clear—do not question leadership, do not think, and do what you are told to do. They serve as an example for all others. This is a cultural condition, and it influences people the wrong way. Think about the impact on everyone. So many people feel numb deep down because true change is not possible with this type of leader and leadership. How many people feel discouraged every day because of how a leader conducts themselves? This type of leader will never have the awareness of their negative impact because of another condition they have created.

Fourth, a condition that weak safety leaders create is the fact that they are not open to much feedback. Feedback creates opportunities for collaboration and learning. Feedback gives people room to make mistakes and learn from those mistakes. Feedback creates an opportunity for trust to grow. However, if this type of leader were open to feedback, they would be forced into awareness of their negative impact, and poor performance. Their ego gets in the way and their pride rules over any healthy vision, purpose, mission, or values. They falsely engineer their results, and therefore it makes everything good for them. Good for them! These types of leaders think telling people what to do works, and when it does not work, they blame others, play politics, and display a lack of character that makes it very difficult for people to function at their full capacity. They have created a condition where people do not report incidents or injuries because of the punitive nature of the culture that has been created.

A *fifth* condition is the safety leader's motivation. There are two basic types of motivators, intrinsic and extrinsic. Intrinsic motivation develops from your personality strengths. It is part of your natural hard wiring. Certain things make you curious and set off chemicals in your brain, and if your curiosity reaches a certain threshold level (i.e., dopamine), you become motivated to learn because you want to, because your personality functions this way.

Extrinsic motivation develops from external factors and is also based on your personality. Depending on the way your personality functions, you are intrigued by something that is outside of you, an external factor. How you are motivated is supported through reinforcement (positive, negative, or recovery), meaning the more reinforcement you receive, the more motivated you will become.

A safety leader's motivations are revealed under stress. They have a tendency when under pressure to leverage their authority, tell people what to do, get poor results, and not accept feedback because of the urgency to achieve results driven by business leaders. This urgency to achieve results (RBS) creates most problems for safety leaders and their organizations. Safety results are an outcome, an indicator that reveals if you are leading correctly. If you motivated correctly, and if you are not getting the results you desire, then you are not leading correctly and need to do better. If you want to do better, you need to realize that there is so much more to leadership than your current limited understanding, and you need to study, learn, and practice how you can influence the conditions and control how they affect you as a safety leader and the people you lead. One area to consider is the alignment between your personality functioning and your natural motivation style (i.e., intrinsic, or extrinsic) and how you are being reinforced.

A healthy condition that safety leaders need to pursue is based on believing in people, which is based on transformational or servant leadership styles. Believing that people come to work to do good work. It is about aligning the vision, purpose, mission, values, competencies, policies, procedures, leading indicators, and technologies into a systematic approach to provide real-time data and pinpointed focus on what matters—having healthy relationships with your people.

Influencing relationships

Influencing relationships is about personal leadership power. Power is defined as (1) the ability to do something or act in a particular way, especially as a faculty or quality. Functioning with power includes ability, capacity, capability, potential, and competence. (2) The capacity, ability, capability, potential, or competence to direct or influence the behavior of others or the course of events.

Influencing relationships reveals your ability to influence or lead people. In a results-based culture these leadership qualities are used negatively. The result is unhealthy relationships and a struggle to sustain results. In a behavior-based culture these qualities are used positively. The result is people are willing to follow you. People are focused on what the safety organization is trying to become. The potential outcome and exponential growth are far greater than the limited results you are trying to achieve. Your leadership is a demonstration of your personal beliefs about people and yourself, what you value about people, and value in yourself. The complexities of leadership are lost in so many leaders because their pursuit of results is like tunnel vision. Results are all they can see, and they miss out on the fact that creating and maintaining healthy relationships is essential to achieving sustainable results. Are you a shallow human being or do you grasp the deeper meaning of healthy relationships? Your beliefs and values are revealed in how you treat people. This is a demonstration of your personal values and your professional values in a safety organization.

At the heart of influencing relationships is communication, connection, inclusivity, support, commitment, boundaries, clearly defined expectations, feeling needed, feeling respected, feeling appreciated, openness, honesty, tolerance, equality, and enjoying the work. All these factors are way in the balance of achieving trust. Healthy relationships are trusting relationships. Achieving trusting relationships is complex and therefore requires a level of wisdom to be successful. The most significant way to achieve healthy trusting relationships is to influence through culture.

Influencing culture

The focus of any safety culture (or corporate culture) should be to bring out the best in people. Bringing out the best in people involves creating conditions in which people are free to perform to their full potential. This is a difficult accomplishment because leadership is not interested in the potential of people, and few people believe in their own potential because they are often conditioned to do what they are told. Because leaders practice *telling* people what to do, a majority of people see themselves as limited. In general, people have a very negative self-regard, and this makes them more susceptible to poor leadership and a dysfunctional culture. Cultural conditions play a significant role in helping people fulfill their purpose and lift them out of this negativity and towards adding value.

Safety leaders are responsible for creating the conditions that shape culture, which are the circumstances affecting the way in which people live or work, especially with regard to their safety or well-being. A safety leader needs to establish an infrastructure that includes organizational design, culture style (mechanistic or organic), cultural phenomena, linking requirements to goals, and integrating expectations with achievement.

Chapter 10

Establishing safety culture infrastructure

The infrastructure of any culture is unseen, but observable. Infra- means "below"; the underpinning of a culture is shaped by vision, purpose, values, behaviors, competencies, processes, and policies that the organization needs in order to function properly. Passenier, et. al. (2016) states that safety culture is often understood as encompassing organizational members' shared attitudes, beliefs, perceptions and values associated with safety. Creating the infrastructure for safety success starts with the right mindset. A mindset is the established set of attitudes held by the safety leader. The mindset is based on attitude, which is defined as a settled way of thinking or feeling about someone and is reflected in the leader's behavior. The right safety leader makes all the difference.

The right safety leader must think and feel appropriately about people. It is the people who matter the most, not your policy or procedure, not your rules, not your overbearing limiting mindset that you have towards people.

The pinnacle of leadership influence is when you can influence the culture. A vast majority of safety leaders do not take the time to focus on culture because they desire short-term results, and only results. The ability to influence a culture is a competency that is rare because business leaders ignore it and expect their safety leaders to ignore it and solely focus on getting results. The focus on results is a blind spot to greater leadership capacity. The fact of the matter is that culture exists if you pay attention to it or not. Understanding and utilizing culture is essential to a successful organization and therefore essential to effective leaders. To create conditions, you have to create infrastructure.

A condition is defined as the circumstances affecting the way in which people live or work, especially regarding their safety or well-being. It also means to bring (something) into the desired state for use. Creating conditions, in the context of the definition, in six specific areas will provide the necessary infrastructure for a healthy culture.

DOI: 10.1201/9781003340799-13

1. Organizational design

This is a methodology which identifies dysfunctional aspects of workflow, procedures, structures, and systems, realigns them to fit current business realities and expectations, and then develops plans to implement the new changes. An organization's structure provides the shared understanding of the accountability and authority of how work is organized and delivered. It defines who does what and who works with whom. Poor design can result in:

- Confusion regarding accountabilities and authorities
- Duplication of effort
- Multiple pulling priorities
- Lack of critical thinking
- Lack of innovation
- Boring, unchallenging, and unsatisfying roles

These poor design results further complicate human performance issues because they create an environment for negative behaviors such as:

- Undermining the work of others
- Constant micromanagement as a leadership technique
- Silo effect
- Group think
- Job protection
- Fear

Such behaviors negatively impact the ability of people to work together and creates a dysfunctional culture that limits the organization and its people. In the organizational culture, the area of conflict tends to be the hand-off points, where work is transferred from one function to the next. Issues often relate to accountability, authority, and resourcing.

In the mechanistic culture, there can be so many levels, and work will become too confined with not enough room for decision-making. There will be overlap and duplication. Authority and accountabilities will likely be blurred, yet everyone is busy. If a work level is missing, there will be a lack of traction in getting action on strategies or plans. Managers will need to dip down to fill the missing work level. If a role is stretched across multiple work levels, its unique value add will be confused or unclear. In each case the outcomes are predictable and will impact working relationships.

2. Establish clearly defined expectations for every employee

This is one of the most overlooked aspects in an organization regarding performance. It is also a significantly caustic condition that is hidden in the safety culture. It creates a variety of symptoms, and the root cause rarely gets diagnosed. All managers must set expectations on how team members are to perform, work together, and achieve their goals—accountability is about having the right conversations, not telling them what to do. Coaching and conversation go together. If expectations are not set, then "the way work gets done around here" will develop in dysfunctional ways limiting the organization because you are limiting the potential of your people.

3. Clearly defined roles and relationships

Clearly defined roles with clear accountabilities provide the rules for engagement. They enable people to work together constructively towards business goals. While the design of all roles is important, one of the biggest sources of negative relationship issues and failure in organizational governance is the design of specialist roles, such as technical specialists and planners. Issues arise when employees do not have a clear understanding of the nature of a specialist's separate, but complementary, work. For example, who sets accountability and authority for the organization's principles and standards? Who is accountable and has the authority to implement, monitor, and report on these principles and practices? What authority do they have and how does it relate to mine? To reduce potential conflict, and improve culture and governance, all roles and relationships must be clearly defined, embedded into the organization's systems of work, and communicated to all those impacted.

4. Provide effective systems of work

Systems of work (e.g., policies, procedures, communications, and IT technologies) coordinate and direct work. They create customs, practices, traditions, beliefs, and assumptions, which in turn create the organization's culture. Systems of work must be designed in a way that supports work and does not hinder it. They must conform to specific design principles such as:

- One system owner
- Measures of performance with leading and lagging indicators
- Feedback mechanisms

When well designed, the influencing systems of work will be highly productive. If poorly designed, their influence will be counter-productive, have poor governance, and will cause conflict.

5. Build strong manager-employee relationships

The foundation of having constructive working relationships is built upon strong manager-employee working relationships based on achieving business goals. Being a managerial leader is not about being charismatic, using charm, trading favors, or relying on working the politics within an organization. Nor is it about building or sustaining personal friendships or social relationships. It is about having the capability to do your role. This is achieved by managers:

- Demonstrating capability in their role
- Providing a safe place to work
- Demonstrating the behaviors of honesty, integrity, and respect
- Consistently and fairly applying the organization's systems of work
- Continually engaging their team by communicating what is required for the business and why

6. Develop emotional and social intelligence

As social interaction is required to achieve business outcomes, the use of good interpersonal skills provides the "social glue" to enable people to work together. Managers need specific skills to support the delivery of their role. These include skills to:

- Effective Performance Safety Coaching©
- Address conflict
- Hold people accountable for unacceptable performance
- Reinforce desired and required behaviors
- Empathize with people's feelings, not apologize
- Recognize quality work

Building a constructive working culture requires an approach that covers the whole enterprise: infrastructure, roles and relationships, systems of work, and managerial leadership, along with the symbols they create. How these are designed and delivered will either enable people to work together constructively or it will hinder them. Focusing on interpersonal skills alone is only a band-aid solution to workplace issues because issues are not resolved so conflict will re-emerge.

A MECHANISTIC OR ORGANIC CULTURE

Any organizational culture can be shaped in one of two ways, and therefore any safety culture can be structured in one of two ways. One, as mechanistic (command-and-control, hierarchical) or two, organic (enabling and learning, flat). Either cultural structure emerges from the leader's leadership style. Everything rises or falls based on leadership. The safety leader has to know how their leadership style aligns with the corporate's culture, and how it impacts the safety culture by knowing how to create or change the culture towards the most effective state of being that produces the most effective environment to support and reinforce defined safety behavior. A mechanistic culture is conducive to Results-Based Safety (RBS) practices, while an organic culture is conducive to Behavior-Based Safety (BBS) practices.

Mechanistic culture

Mechanistic organizations are based on a transactional leadership style. Transactional leadership tends to move towards dictatorial leadership. It begins with the attitude of the transactional leader: (1) people perform their best when the chain of command is clearly defined and adhered to. (2) Rewards and punishments motivate workers, as long as they are willing to play the game. This often comes at a price because it becomes extreme in that rewards become focused on the select few or those in the "inner circle," also known as a click. With rewards also come punishments that are heavily leveraged by the click in unreasonable ways that defy the organization's proposed values, purpose, and vision. (3) Obeying the instructions and commands of the leader is the primary goal of the followers. Total compliance is the "demanded" outcome. There is little to no concern regarding the talents of any individual, or for innovation. Transactional leaders can become outright abusive towards those who do not comply with ridiculous requests that go way below professional level expectations. (4) Subordinates need to be carefully monitored to ensure that expectations are met. People need to be supportively monitored, trusted, and reinforced. Monitoring via software tracking, phone tracking, emails, texting, apps, GPS, security cameras, and hand signatures usually turns into excuses for leadership to punish people because they catch people doing wrong. Transactional leaders seem to enjoy punishing people, it makes them feel important, and thus it is all about them. For some, it is the ultimate ego ride and becomes full-blown arrogance. Authoritarian leaders ride rough over people.

Mechanistic cultures centralize decision-making and establish formal, standardized control systems. Essentially, they are bureaucracies. Mechanistic organizations work well in stable and simple environments. Managers integrate the activities of clearly defined departments through formal channels in formal meetings. Formality tends to become cold and mechanical and de-humanizes relationships by focusing on standards, and constantly raising the bar on goals. A mechanistic organization is characterized by a relatively high degree of job specialization, rigid departmentalization, many layers of management (particularly middle management), narrow spans of control resulting in centralized decision-making, and a long chain of command. Mechanistic cultures focus on policy and procedure, which in turn slows the organization's capacity to achieve its stated goals. Mechanistic leaders consider policy and procedure more important than relationships. Silo effect and clicks plague this type of culture if not careful. Transactional leaders typically foster their own problems so that they can solve them. They often view themselves as problem solvers but conveniently hide the fact that they are the ones who created the problem in the first place.

In a mechanistic culture people are not encouraged to be creative or to find new solutions to problems. Research has found that transactional leadership tends to be most effective in situations where problems are simple and clearly defined. It can also work well in crisis situations where the focus needs to be on accomplishing a specific task. By assigning clearly defined duties to particular individuals, leaders can ensure that those things get done. Transactional leaders focus on the maintenance of the structure of the group. They are tasked with letting group members know exactly what is expected, articulating the rewards of performing tasks well, explaining the consequences of failure, and offering feedback designed to keep workers on task.

While transactional leadership can be useful in some situations, it is considered insufficient in many cases and may prevent leaders, people, and the organization from achieving their full potential. Parkinson's law (Parkinson, 1955) and the Peter principle (Peter, 1970) have been formulated to explain how mechanistic cultures become bureaucracies and suffer from dysfunction.

Organic culture

Organic organizations are more capable and show evidence of more success. The organic culture is a safety leader's best friend because it allows the safety leader the ability to identify, define, and implement conditions based on reinforcement schedules, behavioral modeling practices that are scalable and influence behaviors of individuals, groups, and the culture itself towards a desired or required state of safe behaviors in that environment. In addition to all the technical aspects of safety, the safety leader must understand how people function if they want to achieve their desired outcomes, and the organic culture allows for that to happen. Research indicates that organic organizations provide greater satisfaction for employees and greater levels of self-actualization.

There are challenges associated with flatter structured organic cultures. Research shows that when managers supervise a large number of employees, which is more likely to happen in an organic culture, employees experience greater levels of role ambiguity—the confusion that results from being unsure of what is expected of a worker on the job. This is especially a disadvantage for employees who need closer guidance from their managers. Moreover, in an organic culture, advancement opportunities will be more limited because there are fewer management layers. Finally, while employees report that organic culture is better at satisfying their higher-order needs such as self-actualization, they also report that tall structures are better at satisfying the security needs of employees. Because tall structures are typical of large and well-established companies, it is possible that when working in such organization's employees feel a greater sense of job security.

The degree to which a safety culture is centralized and formalized, the number of levels in the hierarchy, and the type of departmentalization the safety division uses are key elements of a culture's structure. These elements of structure affect the degree to which the company is effective and innovative as well as employee attitudes and behaviors at work. These elements come together to create mechanistic and organic structures. Mechanistic structures are rigid and bureaucratic and help companies achieve efficiency, while organic structures are decentralized and flexible and aid companies in achieving innovativeness and agility towards ever-changing business needs.

Impact of cultural phenomena

Phenomena are a fact that can be observed to study. Phenomena are considered to be something that is impressive or extraordinary. In any safety culture there are a lot of phenomena to be observed. You must know how to recognize healthy phenomena when they occur and capitalize on reinforcing these moments because they are indicators of a healthy culture, or you may be reinforcing unhealthy behavior and creating a sick culture and you do not even realize it. Phenomena are an indicator of the individual's immediate experience, and there are four types of phenomena (i.e., natural, social, visual, and psychological) that impact your safety culture.

Natural phenomena

The first type of phenomena are natural phenomena, and they are manifested based on conditions that are experienced in that culture. Natural phenomena are not because of human interaction but are based on how people experience the organization through their senses rather than by intuition or reasoning. It is a symbiotic relationship, where the environment influences people to do their work. For example, the layout of the buildings, landscaping, or the design of the campus or complex. The colors on the walls, textures of walls (e.g., wood, stone, metal, or glass), decorations of artwork, and the symbols of the organization. The quality of materials and level of technology and sophistication all influence people on an individual level.

The level of quality of equipment, uniforms, technology, traditions of earned certifications, and education levels all influence people. Another aspect of natural phenomena is that they support what gains widespread popularity, what trends, and what people focus on based on what the leadership says, and leadership chooses to reinforce, reward, and recognize as well as punish and penalize.

Social phenomena

The second type of phenomena are social. They occur based on the interactions of smaller groups of people, and with other small groups in that culture. For example, a team of Subject Matter Experts (SMEs) in one department talk among themselves and interact. The same SME group can work with another team of SMEs from another department, talk among themselves, and interact. How small teams get along or pair up to get work done are social phenomena.

There is also the social aspect of conversations in general. How people reach out to each other for help, support, understanding, and the sharing of knowledge. Social phenomena include the unspoken opportunities for people to have permission to join meetings or sit in workshops to learn about different aspects of the company and how it functions. It also includes opportunities to join project teams that interest them and to be welcomed. Integration of the social phenomena into culture is extremely important.

Visual phenomena

The third type of phenomena are visual and observable. Visual in the way the place looks, the architecture, floor plans, offices, office layouts, cubicles, cubicle layouts, outdoor views, break areas, color schemes, and dress code. It produces a vibe and makes you feel a certain way (e.g., professional, formal, casual, sharp, or relaxed) and has an impact on each individual's attitude and perception of the organization. The leadership needs to think about the impact of the visual phenomenon and if it aligns with the desired culture.

The visual phenomena are also experienced in the way leaders model leadership, and peers model behavior. Employees learn from the examples set by those around them. Interactions with the right people, who are aligned with the business culture, strategy, and goals, are key players in helping maintain the level of excellence in an organization and the safety program. Leadership plays a key role, but so do one's peers. Peer modeling is observable every day and is effective in influencing new employees. Bandura (1988) established that peer modeling of good social behaviors passes values easily to another person.

Psychological phenomena

The fourth phenomena are psychological. They are based on individual responses, reactions, interactions, and behavior(s) of those around them, especially the leadership, and how they influence the psychological aspects of that culture. Do people feel safe to speak up, provide ideas, and ask questions? This is known as psychological safety, which is critically important to high performing teams. Psychological safety allows people to show their authentic selves and engage in the work without fear of punitive consequences on one's self-image or status or negative impact on their career. It is also about people trying and failing. Failing at something is a great way to learn. The psychological phenomena should see this practice of failure as courageous and necessary towards innovation and continuous improvement.

Another aspect of the psychological phenomenon is the positivity or negativity of an individual. For example, you may think that one person's negative reaction to a situation or a circumstance has little consequence on the safety culture, but that demonstrates your lack of awareness and its potentially devastating impact on the social and natural phenomena aspects of your culture. In reality, it may not seem like much, but in actuality, your cultural conditions have just changed because of one individual's negativity, and now there is a new level of risk in your safety culture.

Most business and safety leaders do not separate the difference between reality and actuality, and that is a problem. Everyone lives in their own reality, but that does not make it actual. The baseline in any culture is actuality, and it keeps us grounded and focused to pursue what is required to meet the business objectives. Understanding the impact of individual reactions versus responses is one example of how culture experiences small shifts and understanding these shifts is crucial to safety success.

Linking requirements with goals

A requirement is a thing that is needed or wanted. It could also be a thing that is compulsory. Performance requirements are typically ignored because the focus is on goals. Goals are significant in the RBS world because they are the tool used to get results. BBS uses a set of criteria which stipulate how things should perform or the standards that they must achieve in a specific set of circumstances. It requires both positive (P^r) and negative (N^r) reinforcement to shape performance behaviors. This is as opposed to the RBS prescriptive specifications which set out in precise detail how something should be done.

When writing performance requirements, they should be quantifiable and define at minimum, the context and expected throughput, defined specific behaviors, defined competencies, emotional responses (not reactions), demeanor, and attitude. It is imperative that safety leaders create an infrastructure to support and elicit desired performance behaviors. One way to do this is through performance requirements planning.

Performance requirements planning begins by providing a clear definition of expected outcomes for a given position and continues by identifying those behaviors (best practices) and/or competencies that will lead to those outcomes. It defines the cross bar for individual performance and serves to:

- Measure individual performance
- Provide individual performance feedback
- Develop personal improvement and progression plans
- Create and deliver management training/development programs and coaching

At a high level, here are some steps for making this happen:

1. Identify key performance behaviors for a given position. Every competency has between four and six behaviors associated with it.
2. Identify core competencies as well as specific lines of business competencies.
3. Create conditions that reinforce behaviors and competencies, and reward performance outcomes.
4. Define the level or scope of the capability needed by a performer to solve a problem or achieve an objective. A condition or capability that must be met or possessed by a solution or solution component to satisfy a contract, standard, specification, or other formally imposed documents.

Integrating expectations with achievement

Integrating expectations is about inviting a conversation with the employee. It is a situation where a safety leader can have a conversation about what the employee thinks they can do. You present the performance expectations, and what you want them to achieve, and then you have to listen. Seek to understand what you are hearing; seek to understand what it means! Does this person believe in themselves? Does this person have confidence? Knowledge? Skills? Abilities? Attitude? And the Industry Specific Insights for the expectations you are presenting to this individual?

You should be measuring on a scale of 1–5 where this person is at, and the gap between that and where they need to go. These are the areas you will focus on with your coaching—on closing the gaps and helping the individual achieve the expectations. It would also be good to capture this information in the performance review for the employee. It is important that people see the progress they have made. It helps to support and develop the growth of each individual and show them how they have developed their belief in themselves and their confidence in themselves. Ultimately, it's about shaping, supporting, and reinforcing the performance of every individual to achieve the expectations set for them. Using the 1–5 scale is common in organizations. What is not common is a clear performance definition for levels 1–5. The expectations for an individual should be defined at level 3. An individual can achieve level 3 if they work by themselves. Achieving level 4 requires insights from co-workers and training. Self-development from reading and education. Achieving level 5 requires insights from other leaders, learning from challenging situations, and receiving mentoring, but you can't get there alone. The Likert scale that is defined helps management, leadership, and the individual employee realize their growth and development. RBS does not care about the people, but results. However, BBS is all about the growth of people, and results are an indicator if you are effective in the way you are developing and growing your people.

Another aspect of performance that many safety leaders don't understand is discretionary effort. When an employee feels valued for their contributions, they demonstrate discretionary effort. The employee works harder and longer, does what is necessary for success, and does not just do their job description. The safety manager has to set clear expectations for the employee for what they are to produce and provide meaningful feedback and guidance. Set those expectations together. Give them a voice in how they deliver on those expectations, and they will make the effort (discretionary) to go beyond those expectations.

Chapter 11

Create a fully functioning safety culture

ALIGNMENT OF CORPORATE AND SAFETY CULTURE

The success of an organization's culture is based on it being well calibrated to its required outcomes. The CEO must have a simple yet clear vision, but what usually gets lost in all the chaos of new technologies and expectations is the clarity of purpose. It is crucial that the fundamentals of conducting the business of safety are clear: how do people make decisions? What is the process for adopting new behaviors? How are behaviors selected? How are behaviors reinforced? How are commitments agreed upon and accountability maintained? What information gets managed and how is it managed? And how is it connected and aligned together? The executive leadership team needs a comprehensive blueprint that allows for changes to be made effectively and as painlessly as possible. The blueprint must align the organization's strategy in a way that invigorates employees, builds distinctive capabilities in its people, and delivers on the customer experience.

Safety leadership must apply this blueprint effectively by comprehending the organizational factors that slow down its responses currently—that is the way decisions are made, and flow through the various channels of people, and how they are executed. Safety leaders must understand the impact of non-productive meetings, unclear accountabilities, and how the structure of norms is not motivating people. The business and safety leaders must come together and create an infrastructure that addresses key factors successfully. Key factors include the following.

Top talent is attracted to your culture

Talented people are critical to achieving successful performance; however, they are often taken for granted when it comes to organizational design. The psychology of human performance is a significant factor in the design of the organizational infrastructure. Each individual leader's role needs to be shaped within the context of the organization's culture. Personality strengths, capabilities, technical skills, managerial experience, and leadership abilities need to be defined, shaped, applied, and reinforced to ensure

DOI: 10.1201/9781003340799-14

that your top talented people are equipped to foster collaboration and empowered to deliver the required outcomes.

A strong link between the human performance factors and the required performance levels must be created and sustained to achieve sustained success. Defining human performance factors allows business and safety leaders to form teams with defined functionality and purpose which means you can compensate for gaps in proficiency through other team members. You can now balance out performance requirements within a team by selecting people with complementary human performance factors and therefore make the team experience more engaging and successful. Another benefit that is often taken for granted with top leaders is their span of control. Fully functioning teams will enhance the span of control for every leader (i.e., the number of direct reports) and amplify the culture by removing the negatives that people experience and engaging them into a mindset of what is possible.

Supporting accountability for everyone

The organizational design needs to include a purposeful plan that makes it easily understood that people need to be accountable for their part of the work that needs to be done. Business and safety leaders undermine accountability with politics, micromanagement, and a lack of clarity regarding the role of the individual and it stifles the functionality and talent of the person in the role. The ability to execute as an organization is part of the culture and it must be aligned in accordance with the vision, purpose, mission, and values. Accountability is making sure that decision rights are clear and that information can flow rapidly through the proper channels as evidenced by improved execution of strategy. Accountability is a tool that should be utilized by leadership to motivate people in their work.

Accountability needs serious consideration and measurement. Creating a matrix to identify an individual's decision-making can establish the quality and accuracy of clarified decision rights and aligning the motivators to be in harmony with the expected goals of the organization, establishing necessary budget authority, creating decision-making forums, and communication responsibilities must be linked together. Accountability enhances the culture of the organization with the efforts of the safety world, and it establishes individual ownership of delivering to the required expectations. Clarifying accountability draws in isolated groups and engages them to be part of a unified effort. Gradually, people's habits are shaped to follow through on commitments in the context of working together towards the required outcome and in turn makes an organization competitive in a chaotic world.

Every organization has a different style of hierarchy, and it needs to be aligned with the purpose of the organization. The safety world seems to function with a strong pattern of hierarchy no matter the corporate hierarchy style, and this needs to change.

I worked with an organization that had a strong corporate value of "respect for others," and in the corporate setting, it was very evident that this value was practiced. However, when I went out to the field to work with the safety leadership, they were anything but respectful towards others. The F-bomb dropping, angry tone of voice, negative attitudes, and general disgust for the work enhanced a negative culture that was very difficult to work in. What I learned later was that the business leadership was pushing, demanding results by certain deadlines, and applying severe pressure on the field leadership and the workers in a way that safety had become an option to be weighed in the pressure to get things done. The corporate culture was not aligned with the safety culture. The safety culture was mechanistic—a mechanistic culture has centralized decision-making and formal, standardized control systems. They are bureaucracies with strong authority-based leadership that integrates the activities of clearly defined departments through formal channels and in formal meetings.

The function of leadership should be aligned in all areas of the business, and it should reflect the business strategy. Behavior-Based Safety does not fit well in a strong, commanding, top-down hierarchy. The strong commanding leadership breaks down key leadership factors that make Behavior-Based Safety (BBS) work. For example, in a commanding structure leader tells people what to do, but to practice proper BBS format you would ask an individual what they are thinking about a specific safety situation. The style of leadership is completely different. One is authoritarian, and the other is transformational. The culture of the organization must reflect an ability to switch to a coaching style. This is pretty much impossible for a strong commanding style culture. The strong commanding hierarchy also lends itself to a Results-Based Safety (RBS) style because it is a natural fit and the two go hand in hand.

The culture of the organization needs to be aligned with its strategy, and these two aspects should come together in a way that is supportive of the critical capabilities of the people. The culture and strategy supporting its people's capabilities are what distinguish the organization in the marketplace and its impression on the customer. Another aspect to consider is that the mechanistic style culture is very limited in its ability to influence workers to be their best. What happens is that leaders have a small span of control and therefore limited influence. In an organic style culture, the leader expands their influence and therefore expands the span of control. Maximizing the span of control needs to be part of your organizational design, and the organizational strategy needs to support the critical capabilities of your people which has a significant impact on your safety culture. Think through your purpose when designing spans of control, and the layers of your organizational chart. A serious limitation to most mechanistic cultures is the lack of information.

Flattening the organization chart is one thing, and the safety world could use some flattening. But the leadership still must be ensconced in a purposeful and more organic culture where information flows easily and the pressure to get things done is accomplished through working together by creating greater accountability. Reducing layers creates this potential in actuality, but if the structure gets too flat, then your leaders are supervising too much and are not as effective in their ability to influence. When you align the culture with strategy, you can better understand the impact of your leaders, and the actual performance gaps. Those gaps are where you spend your time, supporting, coaching, enhancing your people to the required performance expectations, and executing on the goals of the safety leadership, then the business leadership, and ultimately the CEO's leadership. Safety must be part of the big picture by design. Safety must be the culture by design, not an add on to the culture.

Social reinforcement of safety culture

Social reinforcement occurs when we naturally encounter reinforcement throughout our day. Experiences include professionalism, friendliness, inclusion, acceptance, praise, acclaim, and attention from others. A big part of social reinforcement is self-reinforcement, which occurs when an individual approves of their own behavior based on the defined competencies of the organization. They are able to judge their own actions based on clearly defined behaviors, expectations, and goals they are striving to reach. A healthy culture produces certain outcomes that demonstrate that it is healthy and aligns with scientific theory. For example, social reinforcement and self-reinforcement align well with self-determination theory. Self-determination theory is when an individual has the ability to make choices and manage their own career and life. It is important that a culture reinforce people to feel motivated to take the right actions when they perceive that what they do will have an effect on the outcome.

The scientific underpinnings of social reinforcement are based on several effects. The Pygmalion effect, which is the idea that if people believe that something is true, they will act in a manner consistent with that belief. The Golem effect, which is when negative expectations of an individual cause a decrease in that individual's actual performance. The Galatea effect, which is when high self-expectations result in higher levels of performance, and self-regulation. Self-regulation is a theory that employees can be motivated by monitoring their own progress towards the goals they set and adjusting their behavior to reach those goals.

Considering the critical nature of culture to the full functioning of safety, what does a full functioning safety culture look like? How would you recognize it? The ultimate culture is one that is aligned with scientific protocols and provides social reinforcement and self-reinforcement and drives the motivation of the employees to do their best, enjoy their work, and be proud to work at your organization. However, the underpinning of any safety culture is based on leadership, servant leadership.

CREATING LEADING INDICATORS

Every safety program needs to measure performance. Every safety leader needs to understand the current actuality, and where to make improvements, and therefore shapes the performance of your program. Most safety programs identify and monitor lagging indicators—those that measure results or outcomes that have occurred. Leading indicators support the capacity of the safety leader to predict where support is needed to drive performance. However, it is important to have leading and lagging metrics in place to build an accurate understanding of current and historical performance.

Safety programs with effective performance management in place will include leading indicators. If you think of your safety program like a car, leading indicators look out the windshield and focus on the road ahead, while lagging indicators look backward (out the rear-view mirror) at the road you've already traveled—did you achieve the intended result or not?

Leading indicators should build a specific and a broad understanding of performance and that is critical to the safety leader to see a specific and yet big picture, build insights into patterns and trends, better understand the context of situations that people face, and most importantly know where to focus reinforcement efforts (i.e., frequency and fluency). There are three major buckets of leading indicators that capture crucial data for the safety leader: Operations, Systems, and Behavior. Leading indicators can reveal gaps in performance. Leading indicators are unique to your company and your safety program, and therefore they can be challenging to build, measure, and benchmark, but it is worth the effort and expense. Leading indicators reveal what actions are necessary to achieve your safety goals with measurable outcomes—which are leading indicators themselves on the health of your safety program. Here are some examples of leading indicators based on a leading indicator category, and a leading indicator bucket.

Operations

Risk assessments

Hazard hunt	# of assessments conducted per week, month, or quarter?
Hazard hunt	# of completed JSAs/Hazard Hunts / HAZID
Hazard hunt	Drone Pics of Site (Preventative Actions and Emergency Action Plan)

Hazard hunt	% of assessments completed per week, month, or quarter?
Hazard hunt	Ratio between the levels of risk identified (high, medium, low)
Hazard id/recognition	# of unsafe observations including conditions and behaviors
Hazard id/recognition	# of safe observations including conditions and behaviors
Hazard id/recognition	# of unsafe observations per audit/inspection
Hazard id/recognition	# of unsafe observations reported per employee per time period
Hazard id/recognition	# and % of previously unknown/uncategorized hazards discovered
EHS evaluation	% of system component compliance (Maturity Score)
EHS evaluation	# and frequency of audits performed
EHS evaluation	# of corrective actions
EHS evaluation	# of management system root causes identified by incident investigations
Risk profile	Correlation rate between leading and lagging indicators
Risk profile	# of reviews (to check quality of the process)
Risk profile	# of repeat findings
Risk profile	# of gaps in hazard identification process

Management of change process

% of tasks completed
# of facilities running 10% overtime
# of gaps in management of change review
# of new assessments for changes in processes or equipment
# of new trainings

Preventative actions

Avg. days to close
# of days to completion
% closed on time (within X hrs or by due date)
# of open issues that need to be closed
# of open issues that haven't yet had a corrective action assigned

Systems

General

and % of issues in compliance with recommended corrective actions
% or ratio of corrective actions at each level of control
% or ratio of corrective actions according to hazard type

# of corrective actions prioritized by risk
# of divisional targets that have dropped below a 90% completed rate

Compliance

# of regulatory inspections without findings (i.e., OSHA, EPA, API)
% of defect-free agency inspections

Safety app

# of users
% of users

Learning management system

and % of completed training goals (by individual, group, or facility)
% of compliance versus goal
# of training hours (per employee, per site, per time frame)
# of incidents with a root cause that includes lack of training
# of certified trainers/coaches

Safety perception survey

and frequency of perception surveys
% of employees polled
Response rate to the safety perception survey
% of positive vs. negative results
Employee management gap analysis

Permit to work

# of safety inspections and audits per area
# of gaps in completion (e.g., JSAs, Confined Space Entry, SIMOPS)
or % of supervisors and managers who have completed training in permit to work/Hot Work Permits

Behaviors

Safety leader

# of reinforcements
Rating of communication style (i.e., tone of voice, attitude, energy level)
Rating of leadership style
Rating of leadership competencies
Rating of involvement (i.e., critical, non-critical, necessary, non-necessary)

Leadership engagement

# of employee suggestions implemented by leadership
# of employees volunteering for initiatives

# of managers/supervisors participating in critical design reviews
% of positive ratings of managers/supervisors by employees
# of coaching audits

Employee engagement

% of participation rate
# of on-the-job observations from employees
# of off-the-job observations from employees
# of employees personally engaged by walk-arounds

Performance coaching

% of coached observations
# of coached observations
% of peer-to-peer documented observations
# of peer-to-peer documented observations
# of employees leading safety meetings

Performance coaching audits

Asked open-ended questions to start conversation in order to understand the performer's perspective
Practiced active listening (*30%/70% rule) in order to determine the performer's experienced consequence
Rate the % of listening to talking
Expressed empathy when opportunity was presented in order to connect with the performer
Identified and clarified behavioral root issue in order to establish the desired defined behavior
Gained agreement with the performer on desired defined behavior in order to establish accountability
Asked the performer how they will improve and establish the next steps on what the performer will do to improve
Asked the performer for feedback about the coaching conversation

At risk and safe behaviors

Ratio of positive to negative observations
# of observers
% of supervisors meeting observation goals
Ratio of peer-to-peer observations to supervisory observations

Creating leading indicators and collecting the data allows a safety leader to gain insights they need to have in order to understand what is happening in their safety program within days, if not hours. This is a vast improvement over safety programs that collect no leading indicator data. The leading indicator data provides the safety leader with the capability to analyze trends, recognize gaps in performance, and most importantly the data provides information on where to focus the efforts (reinforcements) to prevent incidents and injuries.

References

Bandura, A. (1977a). Self-efficacy: Toward a unifying theory of behavioral change. *Psychological Review*, 84(2), 191–215.

Bandura, A. (1997b). *Self-efficacy: The exercise of control*. New York: W. H. Freeman.

Bandura, A. (1986). *Social foundations of thought and action: A social cognitive theory*. Englewood Cliffs, NJ: Prentice-Hall.

Bandura, A. (1988). Self-regulation of motivation and action through goal systems. In V. Hamilton, G. H. Bower, & N. H. Frijda (Eds.), *Cognitive perspectives on emotion and motivation* (pp. 37–61). Dordrecht: Kluwer Academic Publishers.

Bandura, A. (1997). *Self-efficacy: The exercise of control*. New York: W. H. Freeman.

Bass, B. M. (1990). From transactional to transformational leadership: Learning to share the vision. *Organizational Dynamics*, 18, 19–32. https://doi.org/10.1016/0090-2616(90)90061-S.

Bureau of Labor Statistics (2020). https://www.bls.gov/news.release/pdf/osh.pdf.

Burns, J. M. (1978). *Leadership*. New York: Harper & Row.

Champoux, J. E. (2011). *Organizational behavior: Integrating individuals, groups, and organizations* (4th ed). New York: Routledge.

Colquitt, J. A., LePine, J. A., & Wesson, M. J. (2015). *Organizational behavior: Improving performance and commitment in the workplace* (4th ed.). New York: McGraw-Hill Education.

Cooper, J. O., Heron, T. E., & Heward, W. L. (2019). *Applied behavior analysis* (3rd ed.). Hoboken, NJ: Pearson Education.

Creating a safety culture (n.d.). OSHA. http://www.osha.gov/SLTC/etools/safetyhealth/mod4_factsheets_culture.html.

Daft, R. L. (2008). *The leadership experience* (4th ed.). Mason, OH: Thomson/South-Western.

Donatelle, R. (2011). *Health: The basics* (Green ed.). San Francisco, CA: Pearson Benjamin Cummings.

Dubrin, A. J. (2010). *Leadership: Research findings, practice, and skills* (6th ed.). Mason, OH: South-Western/Cengage Learning.

Edmondson, A. (1999). Psychological safety and learning behavior in work teams. *Administrative Science Quarterly*, 44(2), 350–383.

Eisenbrey, R. (2013). *Workplace injuries and illnesses cost U.S. $250 annually.* Washington, D.C.: Economic Policy Institute.

Fabrigar, L. R., Petty, R. E., Smith, S. M., & Crites, S. L. (2006). Understanding knowledge effects on attitude-behavior consistency: The role of relevance, complexity, and amount of knowledge. *Journal of Personality and Social Psychology*, 90(4), 556–577.

Festinger, L. (1957). *A theory of cognitive dissonance.* Redwood City, CA: Stanford University Press.

French Jr., J. R. P., & Raven, B. H. (1959). The bases of social power. In D. Cartwright (Ed.), *Studies in social power* (pp. 150–167). Ann Arbor, MI: Institute for Social Research.

Glasman, L. S., & Ablarracín, D. (2006). Forming attitudes that predict future behavior: A meta-analysis of the attitude-behavior relation. *Psychological Bulletin*, 135, 778–822.

Greenleaf, R. K. (1970). *The servant as leader.* Cambridge, MA: Center for Applied Studies.

Hazards of behavior-based safety programs (n.d.). UFCW. Retrieved from http://www.ufcw.org/your_industry/retail/safety_health_news_and_facts/behavior_based.cfm

Howes, J. Behavior-based safety programs (n.d.). UAW Health and Safety Department. Retrieved from http://www.ble272.org/Behaviorbasedsafety.pdf.

Hughes, R. L., Ginnett, R. C., & Curphy, G. J. (2012). *Leadership: Enhancing the lessons of experience* (7th ed.). New York: McGraw-Hill/Irwin.

Jung, C. G. (1971) [1921]. *Psychological types. Collected works of C. G. Jung.* Vol. 6. Translated by Adler, Gerhard & Hull, R. F. C. Princeton, NJ: Princeton University Press.

Kahn, W. A. (1990). Psychological conditions of personal engagement and disengagement at work. *Academy of Management Journal*, 33(4), 692–724.

Kiazad, K., Restubog, S. L. D., Zagenczyk, T. J., Kiewitz, C., & Tang, R. L. (2010). In pursuit of power: The role of authoritarian leadership in the relationship between supervisors' Machiavellianism and subordinates' perceptions of abusive supervisory behavior. *Journal of Research in Personality*, 44(4), 512–519.

Konarski, E. A., Johnson, M. R., Crowell, C. R., & Whitman, T. L. (1981). An alternative approach to reinforcement for applied researchers: Response deprivation. *Behavior Therapy*, 12(5), 653–666. https://doi.org/10.1016/S0005-7894(81)80137-2.

Kotter, J. P. (2001). What leaders really do. *Harvard Business Review*, 79, 85–98.

Liu, L., & Keng, C.-J. (2014). Cognitive dissonance, social comparison, and disseminating untruthful or negative truthful EWOM messages. *Social Behavior and Personality*, 24(6), 979–994.

Maslow, A. H. (1943). A theory of human motivation. *Psychological Review*, 50(4), 370–396. https://doi.org/10.1037/h0054346.

McCance, A. S., Nye, C. D., Wang, L., Jones, K. S., & Chiu, C. (2013). Alleviating the burden of emotional labor: The role of social sharing. *Journal of Management*, 39, 392–415.

Miller, J. G. (2001). Culture and Moral Development. In Matsumoto, D. (Ed.) *The Handbook of Culture and Psychology* (pp. 151–167).

Moore, M., & Tschannen-Moran, B. (2010). *Coaching psychology manual.* Baltimore, MD: Wolters Kluwer/Lippincott Williams & Wilkins.

Northouse, P. G. (2013). *Leadership: Theory and practice* (6th ed.). Thousand Oaks, CA: Sage.

Parkinson, C. N. (1955). Parkinson's law. *The Economist*. London.

Passenier, D., Mols, C., Bím, J., et al. (2016). Modeling safety culture as a socially emergent phenomenon: A case study in aircraft maintenance. *Computational and Mathematical Organization Theory*, 22, 487–520. https://doi.org/10.1007/s10588-016-9212-6.

Pavlov, I. P. (1897/1902). *The work of the digestive glands*. London: Griffin. Pavlov, I. P. (1928).

Peter, L. J., & Hull, R. (1969). *The Peter principle*. New York: William Morrow & Co Inc.

Premack, D. (1959). Toward empirical behavior laws: I. Positive reinforcement. *Psychological Review*, 66(4), 219–233.

Rogers, C. R. (1959). A Theory of Therapy, Personality, and Interpersonal Relationships: As Developed in the Client-Centered Framework. In S. Koch (Ed.), *Psychology: A Study of a Science. Formulations of the Person and the Social Context* (Vol. 3, pp. 184–256). New York: McGraw Hill.

Rojas Tejada, J., Lozano Rojas, O. M., Navas Luque, M., & Pérez Moreno, P. J. (2011). Prejudiced attitude measurement using the Rasch Scale Model. *Psychological Reports*, 109, 553–572.

Rotter, J. B. (1954). *Social learning and clinical psychology*. New York: Prentice-Hall.

Ryan, R. M., & Deci, E. L. (2000). Self-determination theory and the facilitation of intrinsic motivation, social development, and well-being. *American Psychologist*, 55(1), 68–78. https://doi.org/10.1037/0003-066X.55.1.68.

Safety culture: What is at stake. Center for Chemical Process Safety (CCPS). Retrieved from http://www.aiche.org/uploadedFiles/CCPS/Resources/KnowledgeBase/Whats_at_stake_Rev1.pdf.

Salovey, P., & Mayer, J. D. (1990). Emotional intelligence. *Imagination, Cognition, and Personality*, 9, 185–211.

Schaubroeck, J. M., Shen, Y., & Chong, S. (2017). A dual-stage moderated mediation model linking authoritarian leadership to follower outcomes. *Journal of Applied Psychology*, 102(2), 203.

Schleicher, D. J., Watt, J. D., and Greguras, G. J. (2004). Reexamining the job satisfaction-performance relationship: The complexity of attitudes. *Journal of Applied Psychology* 89(1), 165–77.

Schuh, S. C., Zhang, X. A., & Tian, P. (2013). For the good or the bad? Interactive effects of transformational leadership with moral and authoritarian leadership behaviors. *Journal of Business Ethics*, 116(3), 629–640. https://link.springer.com/article/10.1007/s10551-012-1486-0.

Sendjaya, S., & Sarros, J. C. (2002). Servant leadership: Its origin, development, and application in organizations. *Journal of Leadership & Organizational Studies*, 9(2), 57–64.

Skinner, B. F. (1938). *The behavior of organisms: An experimental analysis*. New York: Appleton-Century.

Spears, L. (2010). Character and servant leadership: Ten characteristics of effective, caring leaders. *The Journal of Virtues & Leadership*, 1(1), 25–30.

Thompson, L. (1998). *Personality type: An owner's manuel: A practical guide to understanding yourself and others through typology*. Boulder, CO: Shambhala Publications, Inc.

Thorndike, E. L. (1898). Animal intelligence: An experimental study of the associative processes in animals. *Psychological Monographs: General and Applied*, 2(4), i–109.
Thorndike, E. L. (1905). *The elements of psychology*. New York: A. G. Seiler.
Williams, R. (1999). Cultural safety – What does it mean for our work practice? *Australian and New Zealand Journal of Public Health*, 23(2), 213–214.
Yukl, G. (2006). *Leadership in organizations* (6th ed.). Upper Saddle River, NJ: Pearson/Prentice Hall.
Zhang, Y., & Xie, Y. H. (2017). Authoritarian leadership and extra-role behaviors: A role-perception perspective. *Management and Organization Review*, 13(1), 147–166.

Index

Milton Keynes UK
Ingram Content Group UK Ltd.
UKHW020625121223
434203UK00004B/44